Workplace Drug Testing

Workplace Drug Testing

Edited by
Steven B. Karch, MD, FFFLM

Consultant Pathologist and Toxicologist
Berkeley, California

CRC Press
Taylor & Francis Group
Boca Raton London New York

CRC Press is an imprint of the
Taylor & Francis Group, an informa business

CRC Press
Taylor & Francis Group
6000 Broken Sound Parkway NW, Suite 300
Boca Raton, FL 33487-2742

Visit the Taylor & Francis Web site at
http://www.taylorandfrancis.com

and the CRC Press Web site at
http://www.crcpress.com

Contents

Preface

Workplace drug and alcohol testing is the latest component to be added to the discipline of forensic toxicology, which now comprises the triad fields of post-mortem, human performance, and workplace drug testing toxicology. Since drug testing is conducted as a deterrent to drug abuse, and society seeks to protect the rights of individuals in the workplace, every effort is necessary to prevent harm to persons through false accusation. Drug testing is a multifaceted process. The accuracy and validity of analysis and the use and application of the reported results are concerns. Multiple factors include the methodologies (initial and confirmatory), cutoff concentrations, administrative or legally mandated rules, and the influence of prescribed drugs or environmental exposure.

Acquiring a full understanding of workplace drug testing is necessary for practitioners (toxicologists, physicians, and others). The tenets of forensic science and regulatory requirements influence workplace drug testing. The right questions must be asked and proper caveats applied. Traditionally in medical practice, the test result is only one part of the diagnostic paradigm that also includes history and physical examination. In workplace testing, the test result is the primary and paramount element. There is no diagnostic paradigm and the result must stand alone, evaluated only after the test event by medical personnel identified as medical review officers (MROs).

Workplace testing is employed to ensure safety and productivity in the workplace. It is complex with myriad elements vital to its success. This book examines workplace drug testing, including its background and current status, its regulatory basis, its application in the U.S. and abroad, analytical approaches, the use of urine and other biological matrices, quality control and validity testing, the role of the MRO, and associated legal issues.

The Editor

Steven B. Karch, M.D., FFFLM, received his undergraduate degree from Brown University. He attended graduate school in anatomy and cell biology at Stanford University. He received his medical degree from Tulane University School of Medicine. Dr. Karch did postgraduate training in neuropathology at the Royal London Hospital and in cardiac pathology at Stanford University. For many years he was a consultant cardiac pathologist to San Francisco's Chief Medical Examiner.

In the U.K., Dr. Karch served as a consultant to the Crown and helped prepare the cases against serial murderer Dr. Harold Shipman, who was subsequently convicted of murdering 248 of his patients. He has testified on drug abuse–related matters in courts around the world. He has a special interest in cases of alleged euthanasia, and in episodes where mothers are accused of murdering their children by the transference of drugs, either *in utero* or by breast feeding.

Dr. Karch is the author of nearly 100 papers and book chapters, most of which are concerned with the effects of drug abuse on the heart. He has published seven books. He is currently completing the fourth edition of *Pathology of Drug Abuse*, a widely used textbook. He is also working on a popular history of Napoleon and his doctors.

Dr. Karch is forensic science editor for Humana Press, and he serves on the editorial boards of the *Journal of Cardiovascular Toxicology*, the *Journal of Clinical Forensic Medicine* (London), *Forensic Science, Medicine and Pathology*, and *Clarke's Analysis of Drugs and Poisons*.

Dr. Karch was elected a fellow of the Faculty of Legal and Forensic Medicine, Royal College of Physicians (London) in 2006. He is also a fellow of the American Academy of Forensic Sciences, the Society of Forensic Toxicologists (SOFT), the National Association of Medical Examiners (NAME), the Royal Society of Medicine in London, and the Forensic Science Society of the U.K. He is a member of The International Association of Forensic Toxicologists (TIAFT).

Contributors

Donna M. Bush, Ph.D., DABFT
Drug Testing Team Leader
Division of Workplace Programs
Center for Substance Abuse Prevention
Substance Abuse and Mental Health Services
   Administration
Rockville, Maryland

Yale H. Caplan, Ph.D., DABFT
National Scientific Services
Baltimore, Maryland

Vincent Cirimele, Ph.D.
ChemTox Laboratory
Illkirch, France

Edward J. Cone, Ph.D.
ConeChem Research
Severna Park, Maryland

Kenneth C. Edgell, M.S.
Past Director (2001–2004)
Office of Drug and Alcohol Policy and
   Compliance
U.S. Department of Transportation
Washington, D.C.

Francis M. Esposito, Ph.D.
Health Science Unit
Science and Engineering Group
RTI International
Research Triangle Park, North Carolina

Bruce A. Goldberger, Ph.D.
University of Florida College of Medicine
Gainesville, Florida

Marilyn A. Huestis, Ph.D.
Chemistry and Drug Metabolism Section
Intramural Research Program
National Institute on Drug Abuse
National Institutes of Health
Department of Health and Human Services
Baltimore, Maryland

Daniel S. Isenschmid, Ph.D.
Toxicology Laboratory
Wayne County Medical Examiner's Office
Detroit, Michigan

Pascal Kintz, Pharm.D., Ph.D.
ChemTox Laboratory
Illkirch, France

John M. Mitchell, Ph.D.
Health Science Unit
Science and Engineering Group
RTI International
Research Triangle Park, North Carolina

Anya Pierce, M.Sc., M.B.A.
Toxicology Department
Beaumont Hospital
Dublin, Ireland, U.K.

Angela Sampson-Cone, Ph.D.
ConeChem Research
Severna Park, Maryland

Theodore F. Shults, J.D., M.S.
Chairman
American Association of Medical Review
   Officers
Research Triangle Park, North Carolina

Joseph A. Thomasino, M.D., M.S., FACPM
JAT MRO, Inc.
Jacksonville, Florida

Marion Villain, M.S.
ChemTox Laboratory
Illkirch, France

CHAPTER 1

Introduction: Drugs in the Workplace

Yale H. Caplan, Ph.D., DABFT[1] and Marilyn A. Huestis, Ph.D.[2]*

[1] National Scientific Services, Baltimore, Maryland
[2] Chemistry and Drug Metabolism Section, Intramural Research Program, National Institute on Drug Abuse, National Institutes of Health, Baltimore, Maryland

CONTENTS

Substance abuse has been a mainstay of society for the past few millennia. Epidemics come and go in somewhat predictable cycles and the problem has never been resolved. Over the past century, as this cyclic equilibrium has shifted, the magnitude of the problem has generally intensified. The number of euphoric substances has grown and their enhanced distribution has increased abuse. Technology has increased the potency of euphoric compounds, i.e., the synthesis of natural opium to heroin and the *de novo* synthesis of non-natural substances such as PCP, LSD, MDMA, and others. Cocaine has been concentrated and modified from the coca plant to produce freebase and crack cocaine. Improvement in horticulture techniques has increased the content of marijuana to 10 to 15% tetrahydrocannabinol (THC).

Workplace drug and alcohol testing is the latest component to be added to the discipline of forensic toxicology, which now comprises the triad fields of post-mortem, human performance, and workplace drug testing toxicology. Since drug testing is conducted as a deterrent to drug abuse, and society seeks to protect the rights of individuals in the workplace, every effort is necessary to prevent harm to persons through false accusation. Drug testing is a multifaceted process. The accuracy and validity of analysis and the use and application of the reported results are concerns. Multiple factors include the methodologies (initial and confirmatory), cutoff concentrations, administrative or legally mandated rules, and the influence of prescribed drugs or environmental exposure.

Acquiring a full understanding of workplace drug testing is necessary for practitioners (toxicologists, physicians, and others). The tenets of forensic science and regulatory requirements influence workplace drug testing. The right questions must be asked and proper caveats applied.

---

* Dr. Huestis contributed to this book in her personal capacity. The views expressed are her own and do not necessarily represent the views of the National Institutes of Health or the U.S. government.

Traditionally in medical practice, the test result is only one part of the diagnostic paradigm that also includes history and physical examination. In workplace testing, the test result is the primary and paramount element. There is no diagnostic paradigm and the result must stand alone evaluated only after the test event by medical personnel identified as a medical review officer (MRO).

Workplace testing is employed to ensure safety and productivity in the workplace. It is complex with myriad elements vital to its success. This chapter and those to follow will overview workplace drug testing including its background and current status, its regulatory basis, its application in the U.S. and abroad, analytical approaches, the use of urine and other biological matrices, quality control and validity testing, the role of the MRO, and associated legal issues. Although alcohol testing and on-site (point of collection) testing are practiced in the workplace, they are not included in this chapter.

1.1 HISTORY

Workplace drug testing has grown at a steady rate over the last 20 years. It is recognized that employee/applicant drug testing has become a standard business practice in the U.S. It is likely that almost half of the American workforce will be tested for illegal drugs this year. A more detailed history is provided by Walsh.[1]

The workplace drug-testing phenomenon did not occur overnight, but rather evolved slowly during the 1980s. At the outset, workplace drug testing began in the U.S. military with most of the testing done in military laboratories by military personnel. Testing was highly regimented; however, even within the military programs (Army, Navy, and Air Force), procedures, equipment, and standards varied considerably. As the use of the new immunoassay-based drug-testing technology spread to the private sector in 1982–1983, there were no regulations, no certified laboratories, no standardized procedures, and many of the devices marketed for testing were not cleared by the U.S. Food and Drug Administration. Medical and scientific questions concerning the accuracy and reliability of drug testing were raised continuously by those who opposed testing and often formed the basis of lengthy litigation. As interest in workplace drug testing increased in the public and private sectors, the need for regulations to establish the appropriate science, technology, and practice became obvious.

The beginnings of regulated testing were initiated in 1983 when the National Transportation Safety Board (NTSB) sent a series of specific recommendations to the Secretary of Transportation demanding action in regard to alcohol- and drug-related accidents, particularly in the railroad industry. The report indicated that seven train accidents occurring between June 1982 and May 1983 involved alcohol or other drugs. In response to the NTSB recommendations, the Federal Railroad Administration (FRA) with the assistance of the National Institute on Drug Abuse (NIDA) began to develop the first Department of Transportation (DOT) drug regulations in 1983. However, it was not until early 1986 that legal obstacles were cleared and the rule went into effect and was fully implemented. During the 1983–1986 timeframe, many companies in the oil, chemical, transportation, and nuclear industries voluntarily implemented drug-testing programs. Without standards and recognized procedures, almost every action incurred controversy. Lawsuits and arbitration caseloads mounted rapidly. Reports of laboratory errors in the massive military program raised concerns that the application of this state-of-the-art technology might be premature. Allegations of employees stripped naked and forced to provide specimens in view of other employees were often repeated and added justification for regulations.

In 1986, the federal government became involved in employee drug testing in a significant way. In March 1986, President Reagan's Commission on Organized Crime issued its final report. Among the recommendations were the following:

The President should direct the heads of all Federal agencies to formulate immediately clear policy statements, with implementing guidelines, including suitable drug testing programs, expressing the utter unacceptability of drug abuse by Federal employees. State and local governments and leaders in the private sector should support unequivocally a similar policy that any and all use of drugs is unacceptable. Government contracts should not be awarded to companies that fail to implement drug programs, including suitable drug testing. Government and private sector employers who do not already require drug testing of job applicants and current employees should consider the appropriateness of such a testing program.

NIDA convened a conference in March 1986. The conference was designed to discuss and achieve consensus on drug-testing issues. Prior to the release of the President's Commission on Organized Crime report, the NIDA position advocating testing for critical and sensitive positions was viewed as radical. However, once the recommendation for widespread testing of everyone employed in both the public and private sectors was proposed by the President's Commission, the NIDA position became one of reasonable accommodation. The conference thus focused on prescribing the conditions under which testing could be conducted. After lengthy discussions, consensus was reached on the following points:

1. All individuals tested must be informed.
2. All positive results on an initial screen must be confirmed through the use of an alternate methodology.
3. The confidentiality of test results must be assured.
4. Random screening for drug abuse under a well-defined program is appropriate and legally defensible in certain circumstances.

The consensus reached at this meeting in 1986 on technical, medical, legal, and ethical issues truly served to provide the foundation for the development of the federal regulations that were to evolve over the next decade and continue to evolve. The responsibility for developing technical and scientific guidelines for these drug-testing programs was assigned to the Secretary of Health and Human Services (HHS) and was delegated to NIDA. An informal advisory group produced the initial set of guidelines in a matter of months. On February 19, 1987, HHS Secretary Dr. Otis Bowen issued the required set of technical and scientific guidelines for federal drug-testing programs. Several months later Congress passed a new law (Public Law 100-71 section 503) that set the stage for the widespread regulation of employee drug testing. Enacted on July 7, 1987, the law permitted the President's *Drug Free Federal Workplace* program to go forward only if a number of administrative prerequisites were met. Among the list of required administrative actions was that the Secretary of Health and Human Services publish the HHS technical and scientific guidelines in the *Federal Register* for notice and comment, and to expand the "Guidelines" to include standards for laboratory certification. The NIDA advisory group had been working on the concept of laboratory accreditation since early in 1986 anticipating the eventuality of laboratory certification. This allowed NIDA to revise the "guidelines" quickly and include a proposed scheme of laboratory certification, which was published in the *Federal Register* on August 13, 1987, less than 6 weeks after the passage of the law.

The HHS Guidelines included procedures for collecting urine samples for drug testing, procedures for transmitting the samples to testing laboratories, testing procedures, procedures for evaluating test results, quality control measures applicable to the laboratories, record keeping and reporting requirements, and standards and procedures for HHS certification of drug-testing laboratories. The basic intent of the guidelines was and remains to safeguard the accuracy and integrity of test results and the privacy of individuals who are tested. Following comment and revision, the scientific and technical aspects of the guidelines remained intact as drafted and were published in the *Federal Register* as the "Mandatory Guidelines for Federal Workplace Drug Testing Programs" on April 11, 1988. In July 1988, utilizing the certification standards developed as part of the "Mandatory Guidelines," a National Laboratory Certification Program was implemented by HHS/NIDA and was administered under contract by the Research Triangle Institute. More than 100 laboratories have been certified in this program since 1988 with approximately 50 remaining certified in 2006.

The U.S. DOT published an interim final rule on November 21, 1988 establishing drug-testing procedures applicable to drug testing for transportation employees under six DOT regulations. These six regulations were published on that same date by the Federal Aviation Administration (FAA), Federal Highway Administration (FHWA), Federal Railroad Administration, U.S. Coast Guard, Urban Mass Transportation Administration, and Research and Special Programs Administration. The interim final rule (49 CFR part 40) followed closely the HHS regulation entitled "Mandatory Guidelines for Federal Workplace Drug Testing Programs." DOT issued its final rule on December 1, 1989 with an implementation date of January 2, 1990. These regulations brought the rest of the transportation modes in line with the railroad industry and, in most aspects, standardized the procedures across the industry. The Congress later passed the Omnibus Transportation Employee Drug Testing Act of 1991. This Act was an extremely important piece of legislation that broadly expanded drug testing in the transportation industry. The impetus for the legislation was a very visible subway accident in New York City where the engineer was found to be under the influence of alcohol. The Act required DOT to prescribe regulations within 1 year to expand the existing DOT drug regulations in aviation, rail, highway, and mass transit industries to cover intrastate as well as interstate transportation and to expand the drug-testing program to include alcohol. Final DOT rules were published in the *Federal Register* in February 1994, which continued to incorporate the HHS "Guidelines" and required implementation by January 1, 1995 for large employers (i.e., >50 covered employees) and January 1, 1996 for small employers. Currently the DOT regulations cover more than 12 million transportation workers nationwide.

Drug testing in the workplace has changed considerably over the last 20 years and the changes have improved the program. The development and scope of regulations related to testing have had an important effect not only to improve the accuracy and reliability of employee drug testing but also to establish the credibility of the testing process and the laboratories' capabilities to routinely perform these tests. The stringent laboratory certification standards imposed on forensic drug-testing laboratories have influenced clinical laboratory medicine, with dramatic improvement over the last decade. A real concern is that the federal regulations may have become too rigid, precluding technological advances. The Substance Abuse and Mental Health Services Administration (SAMHSA), which was mandated oversight of workplace drug testing in 1992, has regularly modified regulations and, most recently, proposed new adaptations in technology in a broad sweeping proposal to include the testing of hair, sweat, and oral fluid in addition to urine specimens. It also proposes the use of on-site tests of urine and oral fluid at the collection site, the establishment of instrumented initial test facilities, and changes in operational standards.[2]

1.2 INCIDENCE OF DRUGS IN THE WORKPLACE

Good comprehensive statistics regarding drug use and testing incidence have not been developed. The National Household Survey (2003) shows that 19.5 million people over 12 years of age used drugs during the past month. Of these 54.6% used marijuana, 20.6% marijuana and other drugs, and 24.8% used other drugs. Use was predominately in the 14- to 29-year-old age group as follows: 10.9% (14–15), 19.2% (16–17), 23.3% (18–20), 18.3% (21–25), 13.4% (26–29), and 14.9% (all others).

The only comprehensive compilation of drug test data is published by Quest Diagnostics. Its Drug Testing Index is compiled semiannually. Quest Diagnostics is one of the largest providers of workplace drug tests performing more than 12 million tests annually. Its results are the best available statistical indication of trends in the field. Over the years the drug positivity rates have gone from a high in 1988 (13.6%) when drug-testing programs started to 4.5% in 2004. The number of positives has been relatively consistent since 1997 (5.0 to 4.5%); see Table 1.1. The positivity rates by drug category for the combined workforce for the last 5 years (2000 to 2004) is shown in Table 1.2.

Table 1.1   Annual Positivity Rates for Combined U.S.
Workforce (more than 7.2 million tests from
January to December 2004)

Year	Drug Positive Rate
1988	13.6%
1989	12.7%
1990	11.0%
1991	8.8%
1992	8.8%
1993	8.4%
1994	7.5%
1995	6.7%
1996	5.8%
1997	5.0%
1998	4.8%
1999	4.6%
2000	4.7%
2001	4.6%
2002	4.4%
2003	4.5%
2004	4.5%

Source: Courtesy of Quest Diagnostics.

Table 1.2   Positivity Rates by Drug Category for Combined U.S. Workforce as a
Percentage of All Positives (more than 7.2 million tests from January to
December 2004)

Drug Category	2004	2003	2002	2001	2000
Acid/base	0.13%	0.18%	0.27%	0.24%	0.08%
Amphetamines	10.2%	9.3%	7.1%	5.9%	5.1%
Barbiturates	2.5%	2.5%	2.6%	2.9%	3.2%
Benzodiazepines	4.5%	4.7%	4.5%	4.5%	3.9%
Cocaine	14.7%	14.6%	14.6%	13.9%	14.4%
Marijuana	54.8%	54.9%	57.6%	60.6%	62.8%
Methadone	1.5%	1.4%	1.1%	0.88%	0.82%
Methaqualone	0.00%	0.00%	0.00%	0.00%	0.00%
Opiates	6.2%	6.4%	5.5%	5.8%	5.4%
Oxidizing adulterants (incl. nitrites)	0.09%	0.19%	0.52%	0.54%	0.92%
PCP	0.38%	0.61%	0.58%	0.59%	0.56%
Propoxyphene	4.4%	4.5%	5.1%	3.5%	2.3%
Substituted	0.66%	0.73%	0.68%	0.51%	0.58%

Source: Courtesy of Quest Diagnostics.

Updated revisions and more specific breakdowns of these statistics may be found on the Quest Diagnostics Web site www.questdiagnostics.com.

REFERENCES

1. Walsh, J.M., Development and scope of regulated testing. In *Drug Abuse Handbook,* S.B. Karch, Ed. in Chief. CRC Press, Boca Raton, FL, 1998, 729–736.
2. Department of Health and Human Services, Substance Abuse and Mental Health Services Administration, Proposed Revisions to Mandatory Guidelines for Workplace Drug Testing Programs (69 FR 19673), April 13, 2004.

CHAPTER 2

Overview of the Mandatory Guidelines for Federal Workplace Drug Testing Programs

Donna M. Bush, Ph.D., DABFT
Drug Testing Team Leader, Division of Workplace Programs, Center for Substance Abuse Prevention, Substance Abuse and Mental Health Services Administration, Rockville, Maryland

CONTENTS

2.1 HISTORY

"The Federal Government, as the largest employer in the world, can and should show the way towards achieving drug-free workplaces through a program designed to offer drug users a helping hand." These words are part of President Reagan's Executive Order (EO) Number 12564,[1] issued September 15, 1986, which served to launch the Federal Drug-Free Workplace Program. This EO authorized the Secretary of Health and Human Services (HHS) to promulgate scientific and technical guidelines for drug testing programs, and required agencies to conduct their drug testing programs in accordance with these guidelines once promulgated. This Federal Drug-Free Workplace Program covers approximately 1.8 million federal employees. Of this total number, approximately 400,000 federal employees and job applicants work in health- and safety-sensitive positions identified as Testing Designated Positions, and are subject to urine drug testing.

The Supplemental Appropriations Act of 1987 (Public Law 100-71, Section 503) outlined the general provisions for drug testing programs within the federal sector, and directed the Secretary of the Department of Health and Human Services (DHHS) to set comprehensive standards for all aspects of laboratory drug testing. The authority to develop and promulgate these standards was delegated to the National Institute on Drug Abuse (NIDA), an institute within the Alcohol, Drug Abuse and Mental Health Administration (ADAMHA). Following the ADAMHA Reorganization Act (Public Law No. 102–321) in 1992, the authority for this oversight now resides within the Center for Substance Abuse Prevention (CSAP), Substance Abuse and Mental Health Services Administration (SAMHSA). The Division of Workplace Programs (DWP) in CSAP, SAMHSA, administers and directs the National Laboratory Certification Program (NLCP), which certifies laboratories to perform drug testing in accordance with the "Mandatory Guidelines for Federal Workplace Drug Testing Programs" (Guidelines). These Guidelines were first published by the Secretary of HHS in the *Federal Register* on April 11, 1988,[2] and were revised and published in the *Federal Register* on June 9, 1994[3] and again on November 13, 1998.[4] Another revision was recently published in the *Federal Register* on April 13, 2004, and now includes specific urine specimen validity testing (SVT) requirements.[5] The intent of these Guidelines is to ensure the accuracy, reliability, and forensic supportability of drug and SVT results as well as protect the privacy of individuals (federal employees) who are tested.

Subpart B of these Guidelines sets scientific and technical requirements for drug testing and forms the framework for the NLCP. Subpart C focuses on specific laboratory requirements and certification of laboratories engaged in drug testing for federal agencies. The Guidelines cover requirements in many aspects of analytical testing, standard operating procedures, quality assurance, and personnel qualifications.

Requirements for a comprehensive drug-free workplace model outlined in the Guidelines[1-5] include:

1. A policy that clearly defines the prohibition against illegal drug use and its consequences
2. Employee education about the dangers of drug use
3. Supervisor training concerning their responsibilities in a Drug-Free Workplace Program
4. A helping hand in the form of an Employee Assistance Program for employees who have a drug problem
5. Provisions for identifying employees who are illegal drug users, including drug testing on a controlled and carefully monitored basis

Several different types of drug testing are performed under federal authority. These include job applicant, accident/unsafe practice, reasonable suspicion, follow-up to treatment, random, and voluntary testing.

Under separate authority and public law, the U.S. Departments of Transportation (DOT) also conducts a similar Drug-Free Workplace Program that applies to more than 250,000 regulated industry employers who employ more than 12 million workers. The Nuclear Regulatory Commission conducts a similar Drug-Free Workplace Program that applies to about 104,000 employees working for its licensees. Both of these programs require that any drug testing performed as part of these Drug-Free Workplace Programs be conducted in a laboratory certified by the U.S. DHHS to perform testing in accordance with the scientific and technical requirements in the Guidelines.

In December 1988, the first ten laboratories were certified by DHHS through the NLCP to perform urine drug testing in accordance with the requirements specified in the Guidelines. As of July 2005, there are 49 certified laboratories in the NLCP. The largest number of laboratories certified at any one time was 91 in 1993. Even though there are fewer NLCP-certified laboratories today, the testing capability of the laboratories overall has greatly increased. This reflects the evolution of the business of drug testing, which includes laboratory chains and individual large-scale laboratories that can consistently and accurately test more than 10,000 specimens/day in accordance with the requirements of the mandatory Guidelines. In 2004, more than 6.5 million specimens were tested under federal requirements.

2.2 SPECIMEN COLLECTION

It is important to ensure the integrity, security, and proper identification of a donor's urine specimen. The donor's specimen is normally collected in the privacy of a bathroom stall or other partitioned area. Occasionally, a donor may try to avoid detection of drug use by tampering with, adulterating, or substituting their urine specimen. Precautions taken during the collection process include, but are not limited to, the following:

1. Placing a bluing (dye) agent in the toilet bowl to deter specimen dilution with toilet bowl water
2. Requiring photo identification of the donor to prevent another individual from providing the specimen
3. Requiring the donor to empty his or her pockets and display the items to the collector
4. Requiring the donor to wash his or her hands prior to the collection
5. Collector remaining close to the donor to deter tampering, adulterating, or substituting by the donor
6. Taking the temperature of the specimen within 4 min of collection

A label that is made of tamper-evident material is used to seal the specimen bottle, and a standardized Federal Custody and Control Form (CCF) is used to identify the individuals who handled the specimen, for what purpose, and when they had possession of the specimen.

The entire collection process must be able to withstand the closest scrutiny and challenges to its integrity, especially if a specimen is reported positive for a drug or metabolite, substituted, adulterated, or invalid.

To ensure uniformity among all federal agency and federally regulated workplace drug testing programs, the use of an OMB-approved federal CCF is required. Based on the experiences of using the current federal CCF for the past several years, SAMHSA and DOT initiated a joint effort to develop a new federal CCF that was easier to use and that more accurately reflected both the collection process and how results were reported by the drug testing laboratories. The federal CCF is a five-page carbonless document, with copies distributed to all parties involved in specimen collection, testing, and reporting.

Copy 1 is the Laboratory Copy and accompanies the specimen(s) to the testing laboratory; Copy 2 is the Medical Review Officer (MRO) Copy and is sent directly to the MRO; Copy 3 is the Collector Copy and is retained by the specimen collector for a period of time; Copy 4 is the Employer Copy and is sent to the Agency representative; Copy 5 is given to the donor when the collection is completed.

An image of Copy 1 is shown in Figure 2.1. The entire form, including instructions, may be viewed at http://workplace.samhsa.gov.

Briefly, the Instruction for Completing the Federal Drug Testing Custody and Control Form is as follows:

A. The collector ensures that the name and address of the drug testing laboratory appear on the top of the CCF and the Specimen I.D. number on the top of the CCF matches the Specimen I.D. number on the labels/seals.
B. The collector provides the required information in STEP 1 on the CCF. The collector provides a remark in STEP 2 if the donor refuses to provide his/her SSN or Employee I.D. number.
C. The collector gives a collection container to the donor for providing a specimen.
D. After the donor gives the specimen to the collector, the collector checks the temperature of specimen within 4 minutes and marks the appropriate temperature box in STEP 2 on the CCF. The collector provides a remark if the temperature is outside the acceptable range.

FEDERAL DRUG TESTING CUSTODY AND CONTROL FORM

|||||||||||||||||||||||||||||||||
1234567

SPECIMEN ID NO.   1234567

STEP 1: COMPLETED BY COLLECTOR OR EMPLOYER REPRESENTATIVE

A. Employer Name, Address, I.D. No.	B. MRO Name, Address, Phone and Fax No.

C. Donor SSN or Employee I.D. No.

D. Reason for Test: ☐ Pre-employment   ☐ Random   ☐ Reasonable Suspicion/Cause   ☐ Post Accident
                    ☐ Return to Duty   ☐ Follow-up   ☐ Other (specify)_____

E. Drug Tests to be Performed: ☐ THC, COC, PCP, OPI, AMP   ☐ THC & COC Only   ☐ Other (specify)_____

F. Collection Site Address:

Collector Phone No. _____

Collector Fax No. _____

STEP 2: COMPLETED BY COLLECTOR

Read specimen temperature within 4 minutes. Is temperature between 90° and 100° F? ☐ Yes   ☐ No, Enter Remark | Specimen Collection: ☐ Split ☐ Single ☐ None Provided (Enter Remark) | Observed (Enter Remark)

REMARKS

STEP 3: Collector affixes bottle seal(s) to bottle(s). Collector dates seal(s). Donor initials seal(s). Donor completes STEP 5 on Copy 2 (MRO Copy)

STEP 4: CHAIN OF CUSTODY - INITIATED BY COLLECTOR AND COMPLETED BY LABORATORY

I certify that the specimen given to me by the donor identified in the certification section on Copy 2 of this form was collected, labeled, sealed and released to the Delivery Service noted in accordance with applicable Federal requirements.

X_____   AM PM ►   **SPECIMEN BOTTLE(S) RELEASED TO:**

Signature of Collector   Time of Collection

(PRINT) Collector's Name (First, MI, Last)   Date (Mo./Day/Yr.) ►   Name of Delivery Service Transferring Specimen to Lab

RECEIVED AT LAB: | Primary Specimen Bottle Seal Intact | **SPECIMEN BOTTLE(S) RELEASED TO:**

X_____ ►

Signature of Accessioner   ☐ Yes

(PRINT) Accessioner's Name (First, MI, Last)   Date (Mo./Day/Yr.)   ☐ No, Enter Remark Below

STEP 5a: PRIMARY SPECIMEN TEST RESULTS - COMPLETED BY PRIMARY LABORATORY

☐ NEGATIVE   ☐ POSITIVE for: ☐ MARIJUANA METABOLITE   ☐ CODEINE   ☐ AMPHETAMINE   ☐ ADULTERATED
   ☐ DILUTE   ☐ COCAINE METABOLITE   ☐ MORPHINE   ☐ METHAMPHETAMINE   ☐ SUBSTITUTED
   ☐ REJECTED FOR TESTING   ☐ PCP   ☐ 6-ACETYLMORPHINE   ☐ INVALID RESULT

REMARKS _____

TEST LAB (if different from above) _____

I certify that the specimen identified on this form was examined upon receipt, handled using chain of custody procedures, analyzed, and reported in accordance with applicable Federal requirements.

X_____

Signature of Certifying Scientist   (PRINT) Certifying Scientist's Name (First, MI, Last)   Date (Mo./Day/Yr.)

STEP 5b: SPLIT SPECIMEN TEST RESULTS - (IF TESTED) COMPLETED BY SECONDARY LABORATORY

	☐ RECONFIRMED ☐ FAILED TO RECONFIRM - REASON
Laboratory Name	*I certify that the split specimen identified on this form was examined upon receipt, handled using chain of custody procedures, analyzed, and reported in accordance with applicable Federal requirements.*
Laboratory Address	X_____
	Signature of Certifying Scientist (PRINT) Certifying Scientist's Name (First, MI, Last) Date (Mo./Day/Yr.)

(Right margin, vertical:) OMB No. 0930-0158

(Right margin, vertical:) PRESS HARD - YOU ARE MAKING MULTIPLE COPIES

||||||||||||||||||||||||||||
1234567   A
SPECIMEN ID NO.

(PLACE OVER CAP)   1234567
                       SPECIMEN BOTTLE SEAL

Date (Mo. Day Yr.)   ____/____/____
Donor's Initials

||||||||||||||||||||||||||||
1234567   B (SPLIT)
SPECIMEN ID NO.

(PLACE OVER CAP)   1234567
                       SPECIMEN BOTTLE SEAL

Date (Mo. Day Yr.)   ____/____/____
Donor's Initials

COPY 1 - LABORATORY COPY

Figure 2.1   Copy 1 of federal CCF.

E. The collector checks the split or single specimen collection box. If no specimen is collected, that box is checked and a remark is provided. If it is an observed collection, that box is checked and a remark is provided. If no specimen is collected, Copy 1 is discarded and the remaining copies are distributed as required.

F. The donor watches the collector pouring the specimen from the collection container into the specimen bottle(s), placing the cap(s) on the specimen bottle(s), and affixing the label(s)/seal(s) on the specimen bottle(s).

G. The collector dates the specimen bottle label(s) after they are placed on the specimen bottle(s).

H. Donor initials the specimen bottle label(s) after the label(s) have been placed on the specimen bottle(s).

I. The collector turns to Copy 2 (MRO Copy) and instructs the donor to read the certification statement in STEP 5 and to sign, print name and date, and provide phone numbers and date of birth after reading the certification statement. If the donor refuses to sign the certification statement, the collector provides a remark in STEP 2 on Copy 1.

J. The collector completes STEP 4 (i.e., provides signature, printed name, date, time of collection, and name of delivery service), immediately places the sealed specimen bottle(s) and Copy 1 of the CCF in a leak-proof plastic bag, releases specimen package to the delivery service, and distributes the other copies as required.

DWP publishes a Urine Specimen Collection Handbook for Federal Agency Workplace Drug Testing Programs, available at http://workplace.samhsa.gov, which provides additional guidance to those who will be collecting federal employee urine specimens in accordance with the Guidelines.[6]

2.3 SPECIMEN TESTING

The procedures described in the Guidelines include, but are not limited to, collecting a urine specimen, transporting specimens to the laboratories, drug and validity testing of the specimen, evaluating test results by qualified personnel, specifying quality control measures within the laboratory, specifying record keeping and reporting of laboratory results to an MRO, and standards for certification of drug testing laboratories by SAMHSA.

The cornerstone of the analytical testing requirements specified in the Guidelines is the "two-test" concept: (1) an initial test is performed for each class of drugs tested and a drug(s) for creatinine, pH, and oxidizing adulterants; if an initial test is positive for drugs or outside defined limits for SVT, (2) a confirmatory test using a different chemical principle is performed on a different aliquot of the original specimen. Specifically, the initial test technology requires an immunoassay for drugs and colorimetric test for creatinine, pH, and oxidizing adulterants. The confirmatory testing technology requires gas chromatography/mass spectrometry (GC/MS) for drugs, and potentiometry using a pH meter, multiwavelength spectrophotometry, ion chromatography, atomic absorption spectrophotometry, inductively coupled plasma mass spectrometry, capillary electrophoresis, or GC/MS for SVT.

The initial test cutoffs as published in the Guidelines[5] are as follows:

Drug Class	Cutoff (ng/ml)
Marijuana metabolites	50
Cocaine metabolites	300
Opiate metabolites	2000
Phencyclidine	25
Amphetamines	1000

The confirmatory test cutoffs as published in the Guidelines[5] are as follows:

Drug	Cutoff (ng/ml)
Marijuana metabolite[a]	15
Cocaine metabolite[b]	150
Opiates	
Morphine	2000
Codeine	2000
6-Acetylmorphine[c]	10
Phencyclidine	25
Amphetamines	
Amphetamine	500
Methamphetamine[d]	500

[a] delta-9-Tetrahydrocannabinol-9-carboxylic acid.
[b] Benzoylecgonine.
[c] 6-Acetylmorphine tested when the morphine concentration is greater than or equal to 2000 ng/ml.
[d] Specimen must also contain 200 ng/ml amphetamine.

SVT for federal employee specimens collected under the Guidelines is required as of November 1, 2004. This includes, but is not limited to, determining creatinine concentration, the specific gravity of every specimen for which the creatinine concentration is less than 20 mg/dl, and pH, and performing one or more validity tests for oxidizing adulterants. A much more complete discussion of specimen testing and reporting results can be found in the Guidelines.[6]

As part of an overall quality assurance program, there are three levels of quality control (QC) required of each certified laboratory:

1. Internal open and blind samples constituting 10% of the daily, routine sample workload (these are constructed by the laboratory as part of their daily testing protocols).
2. External open performance testing samples, which are distributed quarterly (these are prepared by the government under contract).
3. Double-blind QC samples, which constitute 1% of the total number of specimens submitted to the laboratory for analysis, not to exceed 100 per quarter; 80% of these samples are negative for drugs.
4. Federal agencies are required to procure and submit these samples from reputable suppliers.

2.4 LABORATORY RESULT REPORTING TO AND REVIEW OF LABORATORY RESULTS BY AN MRO

After accurate and reliable urine drug test results are completed by a SAMHSA-certified laboratory, the Guidelines require these results to be reported to an agency's MRO. As defined in the Guidelines, the MRO is a licensed physician responsible for receiving laboratory results generated by an agency's drug testing program who has knowledge of substance abuse disorders and who has appropriate medical training to interpret and evaluate an individual's positive test result together with the medical records provided to the MRO by the donor, his or her medical history, and any other relevant biomedical information. The MRO must contact the donor when the donor's specimen is reported by the laboratory as drug positive, adulterated, substituted, or invalid, and give the donor the opportunity to discuss the results prior to making a final decision to verify the test result. The donor is given the opportunity to request a retest when his or her specimen is reported as positive, substituted, or adulterated. The retest (i.e., an aliquot of the single specimen collection or Bottle B of a split specimen collection) is performed at a second certified laboratory, with specific procedures applied, depending on the results reported by the first testing laboratory.

A positive test result does not automatically identify an individual as an illegal drug user. The MRO evaluates all relevant medical information provided to him or her by the donor who tested positive. If there was a legitimate, alternative medical explanation for the presence of the drug(s) in the donor's urine, the test result is reported as negative to the employer; if there is no alternative medical explanation for the presence of drug(s) in the donor's urine, the MRO reports the result as positive to the agency/employer. Additional instruction and guidance for evaluating adulterated, substituted, and invalid test results for federal employee specimens are provided in the MRO Manual for Federal Agency Drug Testing Programs.[7]

It is also necessary that negative laboratory results be reviewed by an MRO. Laboratory results for double blind performance test samples (many of which are negative) are reported to the MRO in the same manner as results for donor specimens. In this manner, negative laboratory results are evaluated as part of ongoing quality control programs initiated prior to specimen submission to the laboratory.

2.5 LABORATORY PARTICIPATION IN THE NATIONAL LABORATORY CERTIFICATION PROGRAM

2.5.1 Application

A laboratory applying to become part of the NLCP must complete a comprehensive application form, which reflects in detail each section of the Laboratory Information Checklist and General Laboratory Inspection Reports. Evaluation of this completed application must show that the laboratory is equipped and staffed in a manner to test specimens in compliance with the Guidelines' requirements in order for the laboratory to proceed with the initial certification process.

In essence, the Guidelines promulgate forensic drug testing standards for the evaluation of a specimen provided by a federal employee, on which critical employment decisions will be made. The processes that govern this testing are regulatory in nature, designed to ensure that this testing is accurate, reliable, and forensically supportable.

2.5.2 Performance Testing

As part of the initial certification process, the applicant laboratory must successfully analyze three sets of 25 samples (minimum), in a sequential order. The progress of this phase of the certification process is determined by the successful identification and quantification of analytes by the laboratory. If the first two sets of 25 samples are successfully completed, the third set of 25 samples is scheduled for receipt, accessioning, and analysis during the initial laboratory inspection site visit. As part of the maintenance certification process, each certified laboratory must successfully analyze a set of 25 samples (20 drug related and 5 SVT related) sent on a quarterly basis by the NLCP. There are five different types of samples developed by the NLCP to ensure accurate and reliable analyte identification and quantification for drugs and validity testing:

1. Routine samples, which may contain an analyte specified in the Guidelines, and are screened and confirmed in accordance with established cutoffs
2. Routine negative samples, which may contain a drug analyte specified in the Guidelines, but at a concentration less than or equal to 10% of cutoff

3. Routine plus samples, which may contain an analyte specified in the Guidelines and interfering and/or cross-reacting substances
4. Routine retest samples
5. Retest samples with interfering substances

For details on the evaluation of the performance test results, please refer directly to Section 3.18 of the April 13, 2004 Guidelines.[5]

2.5.3 Laboratory Inspection

The laboratory facility must be inspected and found acceptable in accordance with the conditions stated in the Guidelines and further detailed in the Laboratory Information Checklist, General Laboratory Inspection Report, Computer Systems Report, and Records Audit Report. Inspectors are trained by the NLCP staff (SAMHSA/DWP technical staff and their contractors) in the use of the detailed NLCP Laboratory Information Checklist and Reports, and the NLCP Manual for Laboratories and Inspectors.

Prior to the inspection, the laboratory is required to submit detailed information concerning its operations. This information is provided to the inspectors prior to the actual inspection. In this way, the inspectors become familiar with the laboratory operation prior to arrival. A brief description of each section completed by the laboratory follows:

A. Instructions to Laboratory
B. Laboratory Information — Physical aspects of the laboratory such as location, hours of operation, staffing, specimen testing throughput, and licenses
C. Laboratory Procedures — Type of analytical equipment, calibration procedures, reagent kits, derivatives and ions monitored for each drug analyte, as well as similar information relating to validity testing

In addition, there are questions relating to certification and reporting of results, electronic reporting of results, as well as a description of the Laboratory Information Management System (LIMS).

In the past few years, a number of significant updates have been implemented in the NLCP inspection program. These improvements were the result of a careful review of the program experience over several years and also reflected the reality of market consolidation and growth that significantly increased the number of specimens tested under the Guidelines. This workload increase made it difficult for inspectors to review sufficient non-negative test results (i.e., positive, adulterated, invalid, and substituted) during the scheduled laboratory inspection. To address this issue, the NLCP increased the number of inspectors and hours of inspections for some of the laboratory inspection categories. By increasing the number of inspectors, the percentage of non-negative test results reviewed by each inspection team was enhanced in the larger laboratories. The number of inspectors for Categories I and II remained unchanged. That is, a Category I inspection consisted of two inspectors (one checklist inspector and one data auditor) performing a 2-day inspection, and a Category II inspection consisted of three inspectors (one checklist inspector and two data auditors) performing a 2-day inspection. For the larger laboratories (i.e., Categories III to V) the NLCP increased the number of inspectors, and greatly enhanced the percentage of non-negative test results reviewed by the inspection teams. A Category III inspection now requires a team of four inspectors (two checklist inspectors and two data auditors, rather than the three inspectors previously inspecting this category of laboratory) conducting a 3-day inspection. A Category IV inspection has a team of five inspectors (two checklist inspectors and three data auditors) conducting a 3-day inspection. A new Category V inspection was established for laboratories with large workloads, usually testing more than 2000 regulated specimens per day. A Category V inspection has a team of nine inspectors (three checklist inspectors and six data auditors) conducting a 3-day inspection.

For those laboratories that use corporate LIMS not under the direct day-to-day observation and control of the responsible persons (RPs) of the laboratories that it serves, there will be a special inspection of the LIMS at the facility where the LIMS is located. This special inspection will be a 1-day inspection using two inspectors, with one a LIMS professional. Each corporate LIMS facility will undergo this inspection regardless of the number of laboratories that it serves. This new approach began in January 2005.

Historically, the NLCP has primarily focused the inspection on the procedures of the laboratory. The NLCP has now balanced that focus with an increased examination of the laboratory's forensic product. To accomplish this, a major audit component has been incorporated into NLCP inspections. To facilitate these audits, HHS requires each laboratory to submit a list to the NLCP staff of the non-negative (i.e., drug positive, adulterated, invalid, substituted, and rejected for testing) primary specimens and split specimens reported for a 6-month period prior to an inspection. Specific guidance on the format and information to be included on the list is provided to the laboratories. For each of the following specimen categories, the laboratory must submit a *separate* spreadsheet in a workbook to the NLCP:

1. Positive for delta-9-tetrahydrocannabinol-9-carboxylic acid
2. Positive for benzoylecgonine
3. Positive for opiates (morphine, codeine, and/or 6-acetylmorphine)
4. Positive for amphetamines (amphetamine and/or methamphetamine and *d*-methamphetamine if performed)
5. Positive for phencyclidine
6. Adulterated
7. Invalid test
8. Substituted
9. Rejected for testing

NLCP technical staff direct the laboratory to make available all the batch data and documentation for a selected number of those non-negative and split specimen test results to facilitate review by the inspectors. A specific number of non-negative specimens (NNS) are identified for in-depth review consisting of all analytical test records and chain of custody documentation. The number of NNSs selected for this in-depth information review is determined by the category of the laboratory and is as follows: 40 for Category I, 120 for Category II, 200 for Category III, 310 for Category IV, and 650 for Category V laboratories. Additionally, the same in-depth information review for 40 PT samples is conducted at each laboratory regardless of category.

As a key part of updating the NLCP laboratory inspection system, the inspector cadre was reviewed with the goal of using a smaller core group of inspectors, each of whom has committed to participating in multiple inspections per year. This has significantly enhanced the consistency and uniformity in the NLCP inspection process. NLCP inspectors now are required to perform at least two to three inspections per year, to document active participation in forensic toxicology/workplace drug testing/regulated drug testing, and to attend mandatory NLCP inspector training on an annual basis.

The document previously known as the Laboratory Inspection Checklist was reorganized into two documents, from which two reports are generated by the inspection team. These two reports are the General Laboratory Inspection Report and the Records Audit Report. There were also some significant changes in the roles and configurations of NLCP inspection teams that represent the combined observations of the individual team members. Although each inspector does not complete a separate checklist, all team members tour the laboratory and participate in documenting/verifying any checklist deficiencies.

Inspectors assigned to the roles of "Lead Inspector" and "Inspector" use the General Laboratory Inspection Report to inspect the laboratory's current standard operating procedures (SOP) and forensic operations and may aid in the review of the NLCP PT records, and method validation. The Lead Inspector has the responsibility to finalize and submit the team's summary General

Laboratory Inspection Report to RTI, Research Triangle Park, NC, the contractor currently handling the technical and logistical aspects of the National Laboratory Certification Program for SAM-HSA/HHS. Inspectors assigned to the roles of "Lead Auditor" and "Auditor" use the Data Audit Inspection Report and review a number of non-negative test results (i.e., positive, adulterated, invalid, and substituted), NLCP PT records, and method validation records. The Lead Auditor has the responsibility to finalize and submit the team's summary Data Audit Inspection Report and NNS Review List to the NLCP.

The checklist inspector(s) review and document all aspects of forensic urine drug testing processes and procedures at that laboratory for program review and evaluation for compliance with the minimum standards of the Guidelines. The lead checklist inspector prepares a summary report reflecting the 12 sections of the inspection checklist and one section of the Computer System checklist reviewed during his site visit along with an Inspection Evaluation Summary. A brief description of each section follows:

- D. Chain of Custody — Assesses laboratory practices to verify specimen identity, maintain specimen integrity, secure specimens, and maintain chain of custody during specimen receiving/accessioning, aliquoting, initial drug testing, SVT, confirmation testing, and specimen and aliquot disposal
- E. Accessioning — Assesses laboratory practices to accept or reject specimens, evaluate specimen integrity, handle split specimens, maintain specimen integrity
- F. Security — Assesses laboratory practices to control and document specimen and record access
- G. Quality Control Materials and Reagents — Assess laboratory practices to prepare or procure and verify drug or SVT QC samples, properly identify them, and establish acceptable performance limits
- H. Quality Assurance: Review of QC Results — Assesses laboratory practices to review control results so as to detect assay problems
- I. Equipment and Maintenance — Assess laboratory practices for checking and maintaining all laboratory equipment
- J. Specimen Validity Tests — Assess laboratory practices for handling of aliquots of specimens during validity testing, performing initial and confirmatory testing as required (at a minimum a test for creatinine, pH, and oxidizing adulterant on all specimens), applying appropriate cutoffs, analyzing appropriate QC samples for both initial and confirmatory testing as required
- K. Initial Drug Tests — Assess laboratory practices to analyze specimens with specific immunoassay methods and analyze appropriate QC samples
- L. Confirmatory Drug Tests — Assess laboratory practices to analyze specimens with appropriate GC/MS procedures and analyze appropriate QC samples
- M. Certification and Reporting — Assess laboratory practices to report negative and non-negative results to the MRO both with the federal CCF and electronically if so desired
- N. Standard Operating Procedures, Procedures Manual — Assess laboratory procedure manual for content, comprehensiveness, agreement with day-to-day observed operations of the laboratory, to determine availability to staff as a routine reference, and for any modifications made to reflect changes in current practice in the laboratory
- O. Personnel — Assess qualifications of the RP, scientists who certify the accuracy and reliability of results, and supervisory staff; assess staffing adequacy of these personnel in relationship to the number of specimens analyzed
- P. Laboratory Computer Systems — Assess laboratory policies and procedures for the validation and security of the LIM system, the handling of electronic records and reports, the ability to provide audit trails, monitoring the LIM system, and responding to incidents and providing disaster recovery, as well as documenting the qualifications of the LIMS personnel

The auditor(s) extensively review the laboratory's chain of custody documentation, analytical data, and reported results for non-negative specimens (i.e., positive, adulterated, substituted, invalid, rejected for testing) reported in the defined 6-month period (1 month prior to the last inspection to 1 month prior to the current inspection). The lead auditor prepares a summary report reflecting the four sections of the records audit report along with an Inspection Evaluation Summary and Summary of Issues. A brief description of each section follows:

R. Specimen Records — Review and evaluate specimen records for completion of chain of custody documents, for identity of specimens, calibrators, and controls, for the individuals performing and reviewing the testing, evidence that the certifying scientists who reported the results to the MROs reviewed all appropriate information

S. Method Validation, Periodic Re-Verification — Review and evaluate revised test methods, and both SVT and confirmatory drug assays for periodic re-verification for levels of detection, linearity, and specificity

T. NLCP Performance Test Records — Review and evaluate the NLCP PT records to determine if they support the reported results and if all remediation to PT errors was taken and acceptable

U. Reports — Review and evaluate non-negative specimen reports, both hard copy and electronic, to determine if they are in accordance with NLCP guidance

The laboratory's first (initial) inspection is performed by two NLCP-trained inspectors. Prior to their arrival at the laboratory site, the inspectors are provided copies of the information supplied by the laboratory concerning its operations, its standard operating procedures, and its testing procedures.

The inspectors complete sections of the checklist similar to those completed by the inspection team for a maintenance inspection and submit the completed document to the NLCP. A summary, or critique, is prepared from the report by an individual independent of that laboratory's inspection. The items in the critique are then evaluated for compliance with the minimum requirements of the Guidelines. It is necessary that a laboratory's operation be consistent with good forensic laboratory practice. Once all requirements are met, the laboratory is certified by the Secretary, DHHS, as being able to perform drug testing of federal employees' specimens in compliance with the Guidelines. A letter is sent to the laboratory conveying its certification in the NLCP.

Then, 3 months after its initial certification, the laboratory is again inspected, with a broadened focus, now evaluating both practice and the results reported by the laboratory. A critique developed from the individual checklist and audit reports is developed by an individual independent of that laboratory's inspection. The issues in the critique are then evaluated for compliance with the minimum requirements of the Guidelines. It is necessary that a laboratory's operation be consistent with good forensic laboratory practice. If all requirements are met or there are minor easily correctable deficiencies, the inspection critique is sent to the laboratory. A cover letter may also be included that outlines issues that must be addressed within a defined time frame. After successful completion of this inspection, a 6-month cycle of site inspections begins. The number of inspectors sent to the laboratory for an inspection depends on the resources necessary to adequately evaluate the laboratory's operation. These resources are allocated based on the laboratory's personnel involved in accessioning and certification, the number of specimens performed under the laboratory's certification, and the number of non-negative specimens reported by the laboratory.

During the maintenance phase of a laboratory's certification, if all requirements are met or there are minor easily correctable deficiencies, the inspection critique is sent to the laboratory. A cover letter may also be included that outlines issues that must be addressed within a defined time frame. A laboratory continues its certified status as long as its operation is in compliance with the Guidelines and consistent with good forensic laboratory practice. Since participation in the NLCP is a business decision on the part of a laboratory and is voluntary, a laboratory may choose to withdraw from the NLCP. Upon such voluntary withdrawal from the NLCP, a laboratory must inform its clients that it is no longer certified in the NLCP and cease to advertise itself as an NLCP-certified laboratory.

2.5.4 Suspension of Certification

If significant deficiencies in the laboratory's procedures are found, an evaluation of these deficiencies is performed by the NLCP, the program staff in the DWP, and the Office of the General Counsel. A report is prepared for the Director, DWP. If it is determined that there is imminent harm to the government and its employees, action may be taken by the Secretary to immediately suspend the laboratory's certification to perform drug testing of federal, federally regulated, and private sector specimens tested in accordance with the Guidelines. The period and terms of suspension depend on the facts and circumstances of the suspension and the need to ensure accurate and reliable drug testing of federal employees.

2.5.5   Revocation of Certification

Several factors may be considered by the Secretary in determining whether revocation is necessary. Among these are (1) unsatisfactory performance of employee drug testing, (2) unsatisfactory results of performance testing and/or laboratory inspections, (3) federal drug testing contract violations, (4) conviction for any criminal offense committed incident to operation of the laboratory, and (5) other causes that affect the accuracy and reliability of drug test results from that laboratory.

2.6 CONCLUSION

Illicit drug use and abuse continue to affect safety and security in the American workplace. Data from the 1995 National Household Survey on Drug Abuse[8] reveal that there were 12.8 million current (or past month) users of illicit drugs. Since that time, there has been an increase in drug use and abuse. In 2003, the last year for which this report is available,[9] the number of individuals, 12 or older, indicating current drug use was 19.4 million.

Tragic events serve as examples of how drug abuse in the workplace can affect society and cause long-term environmental and economic consequences. Examples of such tragedies where substance abuse in the workplace was responsible for death and destruction include the 1986 railroad accident in Chase, MD, the 1991 subway accident in New York City, and the 1989 environmental disaster in Prince William Sound, AK, caused by the grounding of the *Exxon Valdez* oil tanker.

REFERENCES

1. Executive Order 12564, Drug-Free Federal Workplace, *Federal Register*, 51(180), 32889–32893, September 15, 1986, available at http://workplace.samhsa.gov.
2. Mandatory Guidelines for Federal Workplace Drug-Testing Programs, *Federal Register*, 53(69), 11970–11989, April 11, 1988, available at http://workplace.samhsa.gov.
3. Mandatory Guidelines for Federal Workplace Drug-Testing Programs, *Federal Register*, 59(110), 29908–29931, June 9, 1994, available at http://workplace.samhsa.gov.
4. Mandatory Guidelines for Federal Workplace Drug-Testing Programs, *Federal Register*, 63(219), 63483–63484, November 14, 1998, available at http://workplace.samhsa.gov.
5. Mandatory Guidelines and Proposed Revisions to Mandatory Guidelines for Federal Workplace Drug-Testing Programs, *Federal Register*, 69(71), 19644–19673, April 13, 2004, available at http://workplace.samhsa.gov.
6. Urine Specimen Collection Handbook for Federal Workplace Drug Testing Programs, Division of Workplace Programs, Center for Substance Abuse Prevention, Substance Abuse and Mental Health Services Administration, U.S. Department of Health and Human Services, available at http://workplace.samhsa.gov.
7. Medical Review Officer Manual for Federal Employee Drug Testing Programs, Division of Workplace Programs, Center for Substance Abuse Prevention, Substance Abuse and Mental Health Services Administration, U.S. Department of Health and Human Services, available at http://workplace.samhsa.gov.

8. National Household Survey on Drug Abuse: Main Findings 1995, Office of Applied Studies, Substance Abuse and Mental Health Services Administration, DHHS Publication Number (SMA) 97-3127, 1997, available at http://www.oas.samhsa.gov.

9. National Survey on Drug Use and Health: Main Findings 2003, Office of Applied Studies, Substance Abuse and Mental Health Services Administration, DHHS Publication Number (SMA) 04-3964, 2004, available at http://www.oas.samhsa.gov.

CHAPTER 3

The U.S. Department of Transportation's Workplace Testing Program

Kenneth C. Edgell, M.S.

Past Director (2001–2004), Office of Drug and Alcohol Policy and Compliance, U.S. Department of Transportation, Washington, D.C.

CONTENTS

The U.S. Department of Transportation (DOT) oversees the largest drug- and alcohol-testing program in the country. The DOT rules affect more than 12 million transportation employees across the U.S. The program also has international impact in that all motor carriers and some railroad workers whose work brings them into the U.S., either from Canada or Mexico, are subject to the same testing requirements as their American counterparts.

The overall responsibility for management and coordination of the DOT program resides with the Office of the Secretary of Transportation (OST), an Executive Cabinet position appointed by the president. Compliance and enforcement within the different transportation industries are the responsibility of the DOT agency that has regulatory authority over the particular industry. Those

DOT agencies are the Federal Aviation Administration (FAA), Federal Motor Carrier Safety Administration (FMCSA), Federal Railroad Administration (FRA), Federal Transit Administration (FTA), and Pipeline and Hazardous Materials Safety Administration (PHMSA).

Safety has been the highest priority for the Secretary of Transportation since Congress established the department in 1966. One of the means used by the Secretary to ensure DOT maintains the highest degree of safety possible is to subject transportation workers to drug and alcohol testing. The workers who are tested have direct impact on the safety of the traveling public or the safety of those potentially affected by the transportation of hazardous products, such as gas and oil pipeline operations. Any worker who has a positive test or other drug and alcohol violation becomes ineligible to continue performing the duties of his or her safety job. In order for the individual to return to a safety job in transportation, that person must satisfactorily complete certain DOT return-to-duty requirements.

Since the outset of the DOT program, there have been a number of legal challenges from those who oppose drug testing. However, because safety is its sole reason for existing, court decisions have allowed the testing to continue without any major setbacks. The Supreme Court, in fact, found a compelling government need "in an industry that is regulated pervasively to ensure safety."[1] So, the program continues. As might be expected, most states that have developed statutes regarding nonregulated testing have referred to the DOT program as the standard of practice they would also follow.

3.1 BACKGROUND

The Department of Transportation drug and alcohol testing rules were an outgrowth of a highly visible and tragic transportation accident that occurred in 1987. The investigation of the accident revealed that employees in safety jobs, regulated by DOT (in this case train crew members), admitted using marijuana and alcohol prior to the accident. In response to the findings of the accident investigation, Senators Earnest Hollings (D-SC) and John Danforth (R-MO) sponsored legislation requiring drug and alcohol testing in the rail, aviation, and trucking industries. DOT, feeling the obvious impact of the accident and the duty to respond to the public, did not wait for the passage of the bill, but instead implemented drug testing under its own authority in 1989.

The action of DOT appeared to have been insightful in the years to follow. In August 1991, another deadly transportation accident occurred. Barely 2 months later, the Hollings-Danforth bill, which had been stymied in Congress for more than 4 years, was passed and signed into law by then-President George H.W. Bush. That legislation, cited as the "Omnibus Transportation Employee Testing Act of 1991" (OTETA),[2] required drug and alcohol testing of employees occupying safety-sensitive jobs in the transportation industries of aviation, trucking, railroads, and mass transit "in the interests of safety." While the gas and oil pipeline industry and the U.S. Coast Guard's (USCG) maritime industry were not mentioned in OTETA, testing within those industries also began, and continues, under the authority granted to DOT as a government agency with regulatory responsibility.

Like most laws, OTETA provided a general overview of what Congress expected to occur. It provided high-level instructions on who would be tested and for what reasons, how the tests would be conducted, and what would happen if and when someone tested positive. With that basic guidance, the statute put the burden of prescribing the remainder of the detail (i.e., the regulations) on the Secretary of Transportation. To fully meet the requirements of OTETA — and develop regulations that were as consistent as possible across all of DOT — the Secretary established the Office of Drug and Alcohol Policy and Compliance (ODAPC) to manage the development effort for workplace drug and alcohol testing.

ODAPC is a "small" office (staffing has never exceeded more than ten people) whose mission is to ensure that the drug and alcohol testing policies and goals of the Secretary are developed and

Table 3.1   Number of Federally Regulated Employers and Employees Subject to Testing

Industry	Government Agency	No. Employers	No. Employees
Aviation	FAA	7,200	525,000
Highway	FMCSA	650,000	10,941,000
Railroad	FRA	650	97,000
Transit	FTA	2,600	250,000
Pipeline	PHMSA	2,450	190,000
Maritime	USCG[5]	12,000	132,000
Totals		674,900	12,135,000

carried out in a consistent, efficient, and effective manner within the transportation industries for the ultimate safety and protection of the traveling public. This is accomplished through program review, compliance evaluation, and the issuance of consistent guidance material for DOT Operating Administrations (OAs) and for their regulated industries. The director of ODAPC is a "political appointee," in that as the administration changes, the director changes; the staff are career employees. Information about this office, along with the most current updates of program documents, can be found at the ODAPC Web site: www.dot.gov/ost/dapc/.

49 CFR Part 40, the "Procedures for Transportation Workplace **Drug** and **Alcohol Testing** Programs,"[3] was developed by ODAPC. With this document, DOT set the standard by which all of its required drug and alcohol testing would be conducted. Such a standard assures that whether the employee is an airline pilot, truck driver, or railroad engineer, his or her drug and alcohol tests are conducted and reviewed using the same procedures. This regulation also sets the criteria that must be met before a person can return to safety-sensitive work after a drug or alcohol violation. Better known simply as "Part 40," this regulation, or rule (the terms are used interchangeably), has become the standard for drug and alcohol testing in the workplaces across the U.S. ODAPC is responsible for providing any authoritative interpretations, if and when they are necessary, on Part 40.

In 1994, also in response to OTETA, six agencies (or "modes" as they are known) within DOT also published testing regulations.[4] These regulations covered who would be subject to drug and alcohol tests, when and why those tests would occur, and what responsibilities the transportation employers would bear in ensuring that the program was implemented properly. By the end of calendar year 2005, the scope of DOT testing covered approximately 12.1 million transportation workers (Table 3.1). It is estimated that out of this total population about 7,000,000 drug tests are conducted each year under DOT authority. This figure represents approximately 20% of the drug tests conducted in this country on an annual basis. The remaining 80% of the testing falls under "non-regulated" status, where companies of their own volition, not due to any government mandate, conduct tests. The number of alcohol tests conducted each year under DOT authority is far less than the number of drug tests; this number could be as little as 1,000,000 tests per year.

3.2 DOT RELATIONSHIP WITH HEALTH AND HUMAN SERVICES

DOT drug testing follows the guidelines established by the Department of Health and Human Services (HHS). HHS was tasked[6] during the Reagan Administration's "War on Drugs" to develop standards for conducting drug tests on federal employees who occupied positions of a safety- or security-related nature. Ultimately, HHS determined which drugs to test for and how laboratories should conduct the tests, including what cutoffs to use. HHS also established specific standards for certifying the laboratories to conduct drug tests for federal agencies. This material is published in an oddly titled regulation called the "Mandatory Guidelines."[7] More detail on the content of the HHS Guidelines can be found elsewhere in this volume (see Chapter 2).

OTETA required DOT to "develop requirements that, for laboratories and testing procedures for controlled substances, incorporate the Department Health and Human Services scientific and technical guidelines dated April 11, 1988, and any amendments to those guidelines, including mandatory guidelines." With that direction, DOT has taken most of the testing requirements contained in the HHS Guidelines, either by reference or by actually repeating the language, and incorporated them into Part 40 for those entities subject to DOT rules. This approach may seem a little repetitious, but it does allow those having to implement the programs not to have to maintain proficiency with two sets of drug-testing rules issued by two government agencies.

The DOT drug-testing program is a laboratory-based urine-testing program, exclusively. While much ado was made of an "alternative specimen" proposal issued by HHS in 2004, which proposed allowing federal agencies to collect and test specimens other than urine (i.e., hair, oral fluid, and sweat), those specimens are not allowed for use under the DOT rules. Neither is any sort of "on-site" test. In fact, the HHS proposal does not apply to DOT.

For the record: Regardless of what conclusion HHS comes to with respect to "alternative specimens," the DOT drug test program will not change until DOT issues a change to its rule — 49 CFR Part 40. Changing any government rule takes time and includes a public notice that seeks public comment. Prior to issuing a final rule, DOT will publish a Notice of Proposed Rulemaking (NPRM). (The NPRM is the document that will solicit public comment.) Complicated rulemaking, which would be the case if additional specimens were required for use by the DOT program, could take several years to finalize. Bottom line: Any utilization of alternative specimens in DOT drug testing is uncertain at this time. It may be an appropriate subject for a future edition of this text.

The remainder of this section provides a general summary of Part 40. For further understanding or more detailed information on a particular subject, especially for the purposes of trying to implement a DOT program, the reader should obtain a copy of the rule. It is written in what the government refers to as "plain language," rather than gobbledygook, a style of writing more reminiscent of past federal offerings. The regulation is divided into functional sections and, while somewhat lengthy, is fairly easy to follow and understand.

3.3 PROGRAM RESPONSIBILITY

The responsibility to assure that drug and alcohol testing is carried out according to the requirements of DOT lies with the transportation employers. Sometimes referred to as an "unfunded mandate," the U.S. Congress, through the OTETA statute, mandated drug and alcohol testing for certain transportation industries, and instructed the Secretary of Transportation to develop the rules under which each industry must abide. However, neither DOT nor any of its agencies provides funding to transportation employers to offset any of the program costs. The DOT rules instruct transportation employers to implement this very comprehensive program and then hold the employers responsible for compliance. Employers are expected to absorb the costs; the benefit is a safer workplace.

Transportation employers are free to contract out any portion of the drug and alcohol testing program functions; however, employers cannot outsource their compliance responsibilities. *Service agents* is the term given to those who contract directly or indirectly with employers to accomplish the tasks set forth in DOT rules. Service agents include collectors, laboratories, medical review officers (MROs), breath alcohol technicians, consortia/third party administrators, and substance abuse professionals.

3.4 SAFETY-SENSITIVE EMPLOYEES

OTETA required each DOT administration to specify those under its regulatory purview who are subject to testing. These individuals occupy positions known as "covered functions," or, more uni-

Table 3.2   Government Agencies and Safety-Sensitive Job Positions

Industry (Mode)	Safety-Sensitive Job Positions
Aviation (FAA)	Flight crew members, flight attendants, flight instructors, air traffic controllers, aircraft dispatchers, maintenance personnel, and screening and ground security coordinator personnel
Highways (FMCSA)	Commercial motor vehicle operators
Railroads (FRA)	Hours-of-service employees (engine, train, and signal services, dispatchers, and operators)
Mass Transit (FTA)	Vehicle operators, dispatchers, mechanics, and safety personnel (carrying firearms)
Oil and Gas Pipeline (PHMSA)	Pipeline operations and maintenance personnel, and emergency response personnel
Maritime (USCG)	Maritime crew members (operating a commercial vessel)

versally, as "safety-sensitive" positions (Table 3.2). A person who occupies a safety-sensitive job and performs its functions, whether on a full-time, part-time, or intermittent basis, is subject to testing.

3.5 REASONS FOR TESTING

The DOT drug-testing program is a deterrence-based program, testing for prohibited use of illegal drugs regardless of when the employee might use the drugs. The DOT alcohol-testing program is more a fitness-for-duty program; testing for prohibited use of a legal substance in and around the time a person is working. Drug tests may be conducted at any time the employee is at work, while alcohol tests are only conducted prior to, during, or just after the performance of safety-sensitive functions. OTETA requires DOT to have specific categories of testing, or "reasons for test." There are six testing categories: pre-employment, post-accident, random, reasonable suspicion, return-to-duty, and follow-up.

Pre-employment tests may be conducted before an applicant is hired or after an offer to hire. These tests must be conducted prior to the actual first-time performance of the individual's safety-sensitive functions. Pre-employment tests are also required when employees transfer to a safety-sensitive position from a non-safety-sensitive position within the same company. All DOT modes require new employees to take (and pass) a pre-employment drug test before they begin work. Pre-employment alcohol tests are authorized, but not mandated by all modes. DOT leaves the decision to conduct pre-employment alcohol tests to each individual transportation employer.

Post-accident testing is conducted after qualifying accidents or where the performance of the employee could have contributed to the accident. Each administration determines what "qualifying" means, with respect to an accident within its particular industry. Obviously, aviation accidents are drastically different from highway accidents. Referencing the rule that corresponds to one of the six administrations would be necessary to determine the criteria for a qualifying accident and subsequent requirements for testing. Contributing to an accident could be seen, for example, in a citation given to a driver by law enforcement after a moving traffic violation. All modes do require that post-accident alcohol tests be conducted within 8 h, and post-accident drug tests be conducted within 32 h of the occurrence of the accident.

Random tests are conducted on a random, unannounced basis. Random testing rates are established in a consistent manner at the beginning of each year, but can be different from mode to mode. Each administration sets the annual random testing rate, one for drugs and one for alcohol, based on the industry's respective random positive rate for the previous year. There are two possible annual rates for random drug testing — 25 or 50%; random alcohol testing has three possible rates of testing — 10, 25, and 50%.

If an industry has a random drug-positive rate of 1% or less for 2 consecutive years, the administrator may reduce the random testing rate to 25% per year for that industry. If the random positive rate is greater than 1% for any year, the random testing rate must move back to 50% per year. Like random drug testing, a random alcohol testing rate is calculated for each industry based on that industry's positive rate for the past 1 to 2 years. If the industry's positive rate is less than 0.5% for 2 consecutive years, the modal administrator can set the resulting alcohol testing rate at 10%. Between 0.5 and 1% for 2 consecutive years, the administrator can set the rate at 25%. When the industry positive rate goes above 1%, the random testing rate is set at 50%. A mode must hold a random positive rate within the above ranges for 2 consecutive years in order for the testing rate to be reduced; a 1-year increase returns the industry to the next higher testing rate. Employees in the pipeline industry, under PHMSA, and employees in the maritime industry, under USCG, are not subject to random alcohol testing. The best explanation for this difference is that neither industry was mentioned in OTETA.

Reasonable suspicion testing is conducted when a supervisor, previously trained in the signs and symptoms of drug abuse and alcohol misuse, observes behavior or appearance that is characteristic of drug or alcohol abuse of the employee. The test must be based on observations that are specific, contemporaneous, and articulable. A rumor of "a big party" over the weekend is not reason enough to conduct a reasonable suspicion test.

Return-to-duty and *follow-up* testing is conducted when an individual, who has a drug or alcohol violation (e.g., positive test), returns to the workplace to resume his or her safety-sensitive work. Initially, a return-to-duty test is given. After passing the return-to-duty test, the employee is eligible to return to safety work. Upon returning to work, the person will be placed back in the company's random testing pool. At the same time, the person must also be subject to unannounced follow-up testing to be conducted at least six times within the first 12 months of returning to work.

3.6 CONSEQUENCES OF A DRUG OR ALCOHOL VIOLATION

All workers committing a violation of the DOT drug or alcohol rules must be removed immediately from their safety-sensitive job and are ineligible to return until they satisfactorily complete the DOT return-to-duty requirements. A DOT drug or alcohol violation includes:

- Alcohol tests with a result of 0.04 or higher alcohol concentration
- Verified positive drug tests
- Refusals to be tested (including verified adulterated or substituted drug test results)
- Other violations of DOT agency drug and alcohol testing regulations (such as using or possessing alcohol or illicit drugs while on duty, or using alcohol within 4 h–8 h for flight crews and flight attendants — of reporting for duty, or using alcohol within 8 h after an accident or prior to a post-accident test being conducted, whichever comes first)

3.7 SPECIMEN COLLECTIONS

DOT has established a specific set of procedures for collecting a urine specimen. Precautions are built into the process to help ensure the control and integrity of the collection. The detail that DOT has devoted to collecting the specimen will serve to minimize the collection errors and maximize the probability that the employer will be able to rely on the test result, regardless of its outcome. DOT has built the model system, but it is the individual diligence of each collector upon which that system relies. Obviously, a problem occurring during the collection has the potential to "ripple" and affect the test outcome.

A DOT urine specimen must be collected using a standard collection kit, documented with a standard form, and conducted by a trained collector. All DOT specimens are collected as "split specimens." The standard kit consists of a single-use collection container that has an attached temperature strip for reading the urine temperature, two sealable plastic bottles for the "split

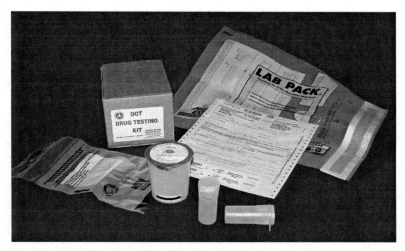

Figure 3.1 Example of DOT Standard Urine Collection Kit.

specimens," a leak-resistant plastic bag with two separate pouches (one for the bottles and the other for the collection paperwork), absorbent material, and a shipping container to protect the specimen during transit to the laboratory (Figure 3.1).

The standard form for all DOT collections is the Federal Drug Testing Custody and Control Form, or CCF. The CCF is a five-part form. DOT requires the same form for collections as is used within the federal program, under the purview of HHS. Figure 2.1 in Chapter 2 presents an illustration of the CCF.

The collector is the person who is in charge of and assists with the collection and has been trained under the provisions of DOT's Part 40 and the DOT Urine Specimen Collection Guidelines,[8] written by ODAPC. Collectors need to be trained prior to collecting their first specimen.

All DOT collections allow the donor to provide the specimen in the privacy of a bathroom stall unless there is a suspicion of tampering, on the part of the donor, or preexisting conditions allow an exception (i.e., return-to-duty and follow-up). The exception would be to conduct the collection under direct-observation procedures, whereby a same-sex collector watches the specimen flow directly from the donor's body and into the collection cup.

DOT has broken down the specimen collection procedure into some 23 different steps in the Collection Guidelines. In general, the donor will present for a urine collection and provide the collector with a form of identification that includes a picture ID. The collector will briefly explain the collection process to the donor emphasizing that, once the collection begins, the donor cannot leave the collection site until excused by the collector. DOT directs collectors to obtain a minimum of 45 ml of urine in one single void; 30 ml is for the primary specimen and 15 ml is for the "split" specimen. Generally collections take less than 15 min to complete.

Donors will be informed that their collection period will be extended up to 3 h in order to obtain the requisite volume of urine should the donor be unable to provide the full 45 ml in a single void. During the additional time period, the donor will be encouraged to drink up to 40 oz. of fluid. After 3 h, if the donor fails to provide the required amount, the collector will inform the donor's employer. The employer must then send the employee for a physical examination. The finding of the examination must be that the individual has a current physical, or a pre-documented psychiatric, condition, or the final result will be deemed a "refusal to test." This expanded collection, with the follow-on physical examination, is referred to as the "shy bladder" process.

During the instruction phase of the collection, the collector will also instruct the donor that the failure to follow any of the collector's instructions could result in the collector stopping the testing process and informing the donor's employer that the donor has "refused to test."

The collection procedures include precautions to guard against possible tampering, such as toilet-water bluing and having donors empty their pockets to reveal any adulterating-type products. Donor are also required to wash their hands before entering the bathroom stall. When the donor presents the collector with the specimen, the collector will examine the specimen for signs of tampering and check the temperature to make sure it is within the acceptable range (90 to 100°F). Any attempt by the donor to adulterate the specimen, which is detected by the collector (e.g., a blue specimen), will result in a second collection to follow immediately. That collection will be conducted using the direct-observation procedures. Should the donor refuse to permit the same-sex, direct-observation collection to occur, the collector will stop the collection and inform the donor that the donor has refused to test. "Refusals to test" are final results of record, and are DOT violations.

Assuming that the collection has gone without incident, the collector will divide, or "split," the specimen into two separate specimen bottles and seal and label both bottles with uniquely numbered tampering-evident labels that are an integral part of the CCF. This is an important step and must be witnessed by the donor. The CCF paperwork is completed with both the donor and collector filling out their specific sections, and then the split specimens and one copy of the CCF are placed in the leak-proof plastic bag, which is sealed and ready to be sent to the laboratory.

It is the responsibility of the collector to secure the specimens until they are sent to the laboratory. The final duty of the collector is to distribute the remaining copies of the CCF to the appropriate parties (i.e., MRO, collector, employer, and employee).

Even though DOT has developed specific training requirements for all collectors, with more than 7,000,000 collections each year, collection errors still occur. This is perhaps inevitable since the position of *collector* is still an entry-level position in most companies and, thus, subject to high turnover. Lack of proper training or training that is administered in a hurried manner is another cause of collection errors. However, any collector who makes an error that results in a test being canceled must be retrained.

3.8 LABORATORY TESTING

DOT makes exclusive use of laboratories certified under the HHS National Laboratory Certification Program. HHS publishes a listing of certified laboratories (each month in the *Federal Register*) of those meeting the criteria set forth in the HHS Guidelines. Currently, the vast majority of these laboratories are located in the U.S.; there are a couple in Canada, but none in Mexico. DOT follows HHS criteria for both drug testing and specimen validity testing. The criteria are specified in detail in Chapter 2 of this text. Therefore, only general references are made to DOT test criteria in this text section.

All laboratories receive, unpackage, and enter a DOT specimen into the testing process. This is called *accessioning*. All DOT specimens are tested for the five drugs of abuse: marijuana, cocaine, opiates, amphetamines, and phencyclidine. The drug panel has not changed since the outset of the program. DOT follows the protocols set up by HHS. Some of the test criteria, such as the cutoffs for initial or confirmation tests, have changed over the years. As testing technology advances and drug-use tendencies change, it is quite possible that similar changes may occur again in the future. It is also possible that additional drugs could be added to the test panel. However, such changes are not made without first being proposed to the public, in order that the public may comment on the recommendations, and then issuing a change to the HHS and DOT rules.

All DOT specimens also undergo "specimen validity" testing. This is a relatively new category of tests brought on in the late 1990s by attempts of individuals to beat the test by tampering with the specimen (e.g., adding a substance to the urine specimen in hopes of altering the test result).

Specimen validity testing (SVT) consists of measuring creatinine and specific gravity to detect a *diluted* or *substituted* specimen. pH is measured as one criterion established to detect an *adulterated* specimen. HHS has developed criteria to be used in testing for specific adulterants such as nitrites, chromates, surfactants, and other active chemical compounds. Substituted or adulterated specimen results are DOT rule violations and have the same weight as a positive test.

Sometimes neither drug testing nor SVT can produce a result that is conclusive. The laboratory may be able to determine only that the specimen has some abnormal reaction. Testing the specimen reveals that it has definitely not met criteria to be reported as negative, however, criteria are not met to call it positive, substituted, or adulterated. This result is classified as *invalid*. Most invalid test results are recollected using directly observed collection procedures.

All results fall into one of two categories: negative and non-negative. *Negative* specimens are those that prove to be negative for drugs and do not have any specimen-validity issues. Negative results may also show that the specimen was *dilute*. (Dilute specimen results are not violations; however, DOT allows employers to have a policy of recollecting specimens from employees who have dilute results.) Approximately 95% of all DOT specimens are negative. Negative results are reached in less (laboratory) time than non-negative results. DOT gives laboratories the authority to report negative results using only computer-generated reports.

Non-negative specimens are all other results — positive, substituted, adulterated, and invalid. Since some of these results will require that the employee be removed from his safety-sensitive job, and possibly terminated, DOT requires more documentation be included with these test reports. The laboratory must also provide a copy of the corresponding CCF, generated at the time of the collection and completed by the certifying laboratory scientist who verifies the laboratory results when the report is released.

DOT test results may only be reported to a physician who will review the drug test. This physician is called the Medical Review Officer. DOT prohibits laboratories from reporting results to anyone other than the MRO (e.g., consortia/third-party administrators, employers).

3.9 MEDICAL REVIEW OFFICER

The MRO has been designated by DOT as the "gatekeeper" of the drug-testing program. "MROs" are physicians (a doctor of medicine or osteopathy) who receive all drug test results and make determinations whether the employee has committed a drug violation (e.g., verified positive test result), while maintaining the confidentiality for the employee during an interview process. MROs are required to have knowledge and clinical experience in substance abuse disorders, to be trained on Part 40, and to pass a written examination. Chasing paperwork, listening to far-fetched stories as to why a person was positive or how the specimen became adulterated, and dealing with employers anxious to put people to work are all part of the MRO's duties.

Laboratories *certify* drug test results, while MROs *verify* drug test results. The MRO receives all DOT drug test results directly from the laboratory. Employers or third-party administrators are not authorized to receive the drug test results or act as an intermediary to the MRO. Staff, under the supervision of the MRO, may assist with paperwork duties, establishing donor contact and reporting results to employers. However, only the MRO can conduct the interview with the donor to determine if there is a legitimate medical explanation for the positive, substituted, or adulterated test result.

"Legitimate medical explanations," while rare for these test results, are possible. Recognizing what is acceptable and what is not is subject-matter training for MROs in the DOT-required MRO training courses. Generally, such explanations are limited to prescriptions or medical procedures where drugs are introduced to the donor and can subsequently be verified by the MRO. Additionally, special studies may be set up to prove that the donor can naturally produce a substituted or adulterated urine specimen. To date, there has been one such case for the former (substituted specimen), but none for the latter (adulterated specimen). In fact, as of this writing, there is no known adulterant, introduced *in vivo*, that can interfere with a laboratory's analytical procedure. All adulterants that have an effect on the analytical process or can be detected by laboratory analysis are introduced into the specimen cup — by the donor — *in vitro*.

When the MRO gets a result that is positive, substituted, adulterated, or invalid, the donor will be contacted and interviewed by telephone. Through the interview, if the MRO determines that there is a legitimate medical reason for the donor's test result, the MRO has the discretion to "downgrade" the result (from positive) to negative and forward that result to the employer. The process devised by DOT gives the MRO latitude to downgrade the final result to negative and still provide a safety warning to the employer if the donor is using a medication that would medically disqualify the donor under agency rules or where continued use would pose a safety risk, even though the medication was obtained with a valid prescription.

For those individuals for whom the MRO determines that the reported result will be positive, substituted, or adulterated, the MRO will also provide the donor with the opportunity of having the split specimen tested at another HHS-certified laboratory. However, the employer will imme-diately remove the donor from the safety job being performed on the initial report by the MRO. If the split specimen result does not confirm the initial specimen result, the result reported to the employer will be canceled by the MRO — as if the test never occurred — and the donor will undergo a second collection where the process starts over from scratch. If the split confirms the initial result, which is what normally occurs, the employer and the donor are so notified.

MROs may also be called upon for consultation during the post-violation assessment process that the donor must undergo in order to return to the transportation workplace. MROs are encouraged to cooperate with this part of the process, which usually involves providing only drug quantitations that may be helpful during the assessment. MROs may also be asked to assist the donor in obtaining test records from the laboratory that conducted the testing. DOT requires laboratories to interface with MROs, not donors. Donors are entitled to any records produced as a result of their drug test, and having the MRO act as the intermediary keeps it simple. Likewise, donors have access to MRO records, pertaining to their test, as well as laboratory records.

3.10 ALCOHOL TESTING

OTETA mandated that alcohol tests be part of the dual-testing program. For the first time in the workplace setting, alcohol tests would be conducted alongside drug tests. This new territory presented especially difficult challenges for the DOT.

Alcohol, unlike the drugs that DOT tests for, is a legal substance. While it could not be tolerated in the safety-sensitive workplace, parameters had to be established for off-duty use occurring near the time when the employee would report for duty. Additionally, consideration needed to be given to the period after accidents, but prior to a test being conducted.

Originally the DOT interpreted OTETA language as requiring alcohol testing for all reasons-for-test: pre-employment, random, reasonable suspicion, post-accident, return-to-duty, and follow-up. The 1994 testing rules issued by DOT included requirements for all six test categories. Even-tually, the 4th Circuit Court of Appeals ruled that pre-employment testing was problematic as a routine, mandated test. It was subsequently reverted back to DOT for resolution. DOT decided to "authorize" employers to conduct pre-employment alcohol tests as a condition of employment, and definitively modified the rules in 2001. Today, alcohol testing is authorized for pre-employment and mandated for other test reasons (except random testing) in the same manner as drug tests. Random alcohol tests are not required under PHMSA and USCG rules.

Personnel who conduct alcohol tests are called screening test technicians (STT) and breath alcohol technicians (BAT). Most of the tests are conducted by "BATs," as they are called. Both technicians need to be trained and only use instruments for the test that have been approved by DOT. All alcohol test devices approved for DOT use are certified by DOT's National Highway Traffic Safety Administration (NHTSA) and placed on a use approved listing,[9] which is published in the *Federal Register.*

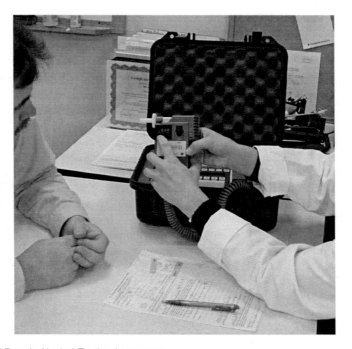

Figure 3.2   DOT Breath Alcohol Testing in process.

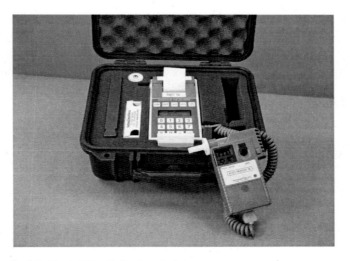

Figure 3.3   Example of Evidential Breath Testing device.

Alcohol tests (Figure 3.2), like drug tests, use a standard form and a specific set of procedures,[10] both designed and developed by DOT. Unlike drug testing, when alcohol was mandated by OTETA, DOT did not have the luxury of following HHS guidelines — no other branch of the government tests for alcohol. Procedures for workplace alcohol testing had to be developed by DOT.

Alcohol tests, like drug tests, use a two-test procedure. The first test, or initial test, can use either breath or saliva testing devices, and may be non-evidential in nature. The second test can only use an Evidential Breath Testing (EBT) device (Figure 3.3). This device produces a printed record documenting all aspects of the test (e.g., time of test, specific device, breath alcohol concentration). Any individual who shows the presence of alcohol at a concentration of 0.02

or higher on the initial test must be subjected to a second test, or confirmation test. Prior to conducting the second test, the BAT will wait at least 15 min, but not more than 30 min, before proceeding. This 15-min waiting period allows any alcohol that may be in the person's mouth to dissipate. "Mouth alcohol" can be attributed to mouthwash or breath lozenges. The waiting period is precautionary to protect the person being tested. During this wait period, the BAT will also conduct an "air blank" on the EBT to show that the device does not contain any residual alcohol.

The result of the second test becomes the result of record. An alcohol concentration of 0.04 or higher is a DOT violation (e.g., similar to a positive drug test); an alcohol concentration between 0.02 and 0.039 is prohibited conduct, and the person cannot perform safety-sensitive work while testing within this range.

The services of an MRO, to review an alcohol test, are not required. The reason is that there is no legitimate medical explanation for alcohol in one's system. Alcohol is alcohol. Whether its source was from beer, whiskey, or mouthwash, the effect is still the same. The BAT will report the result immediately to the employer and the employer will respond by removing the person from duty.

3.11 SUBSTANCE ABUSE PROFESSIONAL

OTETA requires that an opportunity for treatment be made available to covered employees. In order to implement this mandate, an employer must refer any transportation worker who has engaged in conduct prohibited by DOT drug or alcohol rules for evaluation and treatment. Modeled with the substance abuse profession in mind, the DOT rules require substance abuse professionals, or SAPs, to have certain credentials, possess specific knowledge, receive training, and achieve a passing score on an examination. There is also a continuing education requirement. The SAP Guidelines,[11] written by ODAPC, supplement Part 40.

The primary safety objective of the DOT rules is to prevent, through deterrence and detection, alcohol and controlled substance users from performing transportation industry safety-sensitive functions. The SAP is responsible for several duties important to the evaluation, referral, and treatment of employees identified through breath and urinalysis testing as positive for alcohol and controlled substance use, or who refuse to be tested, or who have violated other provisions of the DOT rules.

A SAP's fundamental responsibility is to provide a comprehensive face-to-face assessment and clinical evaluation to determine what level of assistance the employee needs in resolving problems associated with alcohol use or prohibited drug use. Following the evaluation, education and/or treatment is recommended, whereby the employee must demonstrate successful compliance in order to return to DOT safety-sensitive duty.

Prior to the employee's return to safety-sensitive duties, the SAP conducts a face-to-face follow-up evaluation with the employee to determine if the individual has demonstrated successful compliance with recommendations of the initial evaluation. This evaluation must be accomplished before an employer can consider the employee for return to safety-sensitive functions. Therefore, the evaluation serves to provide the employer with assurance that the employee has made appropriate clinical progress sufficient to return to work.

The SAP also develops and directs a follow-up testing plan for the employee returning to work following successful compliance, specifying the number and frequency of unannounced follow-up tests. If polysubstance use has been indicated, the follow-up testing plan should include testing for drugs as well as alcohol, even though a violation of both was not the original offense.

If the MRO is considered the "gatekeeper" in the drug-testing process, then the metaphor is equally appropriate for the SAP in the return-to-duty process.

3.12 CONFIDENTIALITY AND RELEASE OF INFORMATION

Part 40 is very clear about what information is to be generated, where it goes, and who is authorized to receive it. Beyond the Part 40 instruction, service agents and employers participating in the DOT drug or alcohol testing process are prohibited from releasing individual test results or medical information about an employee to third parties without the employee's specific written consent.

A "third party" is any person or organization that the rule (Part 40) does not explicitly authorize or require the transmission of information from in the course of the drug or alcohol testing process. "Specific written consent" means a statement signed by the employee agreeing to the release of a particular piece of information to a particular, explicitly identified, person or organization at a particular time. "Blanket releases," in which an employee agrees to a release of a category of information (e.g., all test results) or to release information to a category of parties (e.g., other employers who are members of a C/TPA, companies to which the employee may apply for employment), are prohibited by DOT.

Information pertaining to an employee's drug or alcohol test may be released in certain legal proceedings without the employee's consent. These include a lawsuit (e.g., a wrongful discharge action), grievance (e.g., an arbitration concerning disciplinary action taken by the employer), or administrative proceeding (e.g., an unemployment compensation hearing) brought by, or on behalf of, an employee and resulting from a positive DOT drug or alcohol test or a refusal to test. Also included are criminal or civil actions resulting from an employee's performance of safety-sensitive duties, in which a court of competent jurisdiction determines that the drug or alcohol test information sought is relevant to the case and issues an order directing the employer to produce the information. For example, in personal injury litigation following a truck or bus collision, the court could determine that a post-accident drug test result of an employee is relevant to determining whether the driver or the driver's employer was negligent. The employer is authorized to respond to the court's order to produce the records.

DOT instructs employers and service agents to notify the employee in writing of any information released without the employee's written consent.

Lastly, release of DOT program information is not in conflict with the HHS Health Insurance Portability and Accountability (HIPAA) rules.[12] HIPAA rules do not conflict with the DOT drug and alcohol-testing program for employer and service agent responsibilities under Part 40 and operating administration drug and alcohol testing rules. Information may be generated and flow to its intended parties, as required by Part 40, without the employee signing a HIPAA-type consent.

3.13 CONSORTIUM/THIRD-PARTY ADMINISTRATORS

In the early days of the DOT testing program, the department believed that employers might pool their resources or band together to help each other accomplish the things the testing rules required. This did not happen. What did happen was that the DOT program created an entire "cottage industry" of *service agents*, including consortia/third-party administrators (C/TPA), and substance abuse providers, who contract directly or indirectly with employers to accomplish the tasks set forth in DOT rules. In short, a C/TPA will handle all of the administrative processes for an employer's testing program. The types of services typically offered by C/TPAs are:

- Urine collections, including permanent and mobile facilities
- Laboratory testing
- Random selections
- Background checks on new hires
- Policy and procedures

- Supervisory training
- Providing MRO test review and SAP referrals
- Maintaining records and preparing statistical data
- Assisting employers preparing for DOT audits

Employers are allowed to contract with C/TPAs (service agents) to get the job done. Service agents are obligated to do the work they have signed on to do as if they were the employer. However, even though the department has a method to rid the testing community of a "bad" service agent, the employer is still held responsible for the actions of the service agent.

3.14 PUBLIC INTEREST EXCLUSION

Service agents perform the bulk of drug and alcohol testing services for transportation employers. Employers, particularly small employers, necessarily rely on service agents to comply with their testing obligations. Employers are ultimately responsible for all aspects of the program. However, in good faith, they may hire a service agent who purposely does not comply with the rules. These employers often do not have the expertise in testing matters that would enable them to evaluate independently the quality, or even the regulatory compliance, of the work that service agents perform for them. Subpart R of Part 40 provides a mechanism to help ensure that service agents will be held accountable for serious noncompliance with DOT rules. The *public interest exclusion* (PIE) is based in concept on the existing DOT non-procurement suspension and debarment rule[13] and permits DOT to suspend a service agent for willful noncompliance with the drug and alcohol testing rules. The mechanism, both for policy and resource reasons, is only used in cases of serious misconduct.

An employer's compliance with DOT regulations is largely dependent on its service agents' performance. If a service agent makes a serious mistake that results in the employer being out of compliance with a DOT rule, accountability must be addressed. The employer may be subject to civil penalties from a DOT agency. The employer can be subject to litigation resulting from personnel action it took on the basis of the service agent's noncomplying services. Most importantly, the employer's efforts to ensure the safety of its operations may be damaged, e.g., as when an employee who apparently uses drugs is returned to duty because of a service agent's noncompliance.

The standard of proof in a PIE proceeding is "the preponderance of the evidence." There is no policy or legal basis apparent for raising this burden to the higher "clear and convincing evidence" level. A PIE could apply to all the divisions, organizational elements, and types of services provided by the service agent involved, unless the director limited the scope of the proceeding. Under some circumstances, affiliates and individuals could also be subject to a PIE. The intent of the PIE is to protect the public from the misconduct of an organization. As of mid-2006, the DOT had not suspended any service agents.

3.15 CONCLUSION

The history of the DOT drug- and alcohol-testing program is a relatively short one. The benefits of the last 10-plus years of program implementation are not fully known. However, data on illegal drug use by transportation employees, accidents related to use, and changes in attitudes about workplace safety and substance abuse are encouraging. The ultimate success of OTETA and the employer-based programs that it mandates will be measured over time in terms of lives saved, injuries prevented, and property losses reduced. OTETA is not a cure-all for safety problems or problem workers. Only everyday due diligence on the part of all safety workers will help in that area.

NOTES

1. *Skinner v. Railway Executives' Association*, 489 U.S. 616-617 (1989); and *National Treasury Employees Union v. Von Robb,* 489 U.S. 656, 674-675 (1989).

2. Public Law 102-143, October 28, 1991, Title V — Omnibus Transportation Employee Testing, 105 Stat. 952-965; 49 U.S.C. 45104(2).

3. Title 49, Code of Federal Regulations (CFR), Part 40, Procedures for Transportation Workplace Drug and Alcohol Testing Programs, Office of the Secretary of Transportation, Department of Transportation.

4. Federal Aviation Administration (FAA): 14 CFR Part 121, Appendix I & J; Federal Motor Carrier Safety Administration (FMCSA): 49 CFR Part 382; Federal Railroad Administration (FRA): 49 CFR Part 219; Federal Transit Administration (FTA): 49 CFR Part 655; Pipeline and Hazardous Materials Safety Administration (PHMSA): 49 CFR Part 199; and U.S. Coast Guard (USCG); 46 CFR Part 16 and 46 CFR Part 4.

5. The USCG transferred to the Department of Homeland Security in March 2003, but still conducts drug testing under DOT rules.

6. Executive Order 12564, Drug-Free Federal Workplace, September 15, 1986.

7. Mandatory Guidelines for Federal Workplace Drug Testing Programs, Division of Workplace Programs (DWP), Substance Abuse and Mental Health Services Administration (SAMHSA), Department of Health and Human Services (HHS).

8. Urine Specimen Collection Guidelines, Office of Drug and Alcohol Policy and Compliance, DOT, Version 1.01, August 2001.

9. Conforming Products Listing, National Highway Traffic Safety Administration, 69 FR 42237.

10. DOT Model Course for Breath Alcohol Technicians and Screening Test Technicians, Office of Drug and Alcohol Policy and Compliance, DOT, August 2001.

11. Substance Abuse Professional Guidelines, Office of Drug and Alcohol Policy and Compliance, DOT, August 2001.

12. Health Insurance Portability and Accountability Act of 1996 (HIPAA), Department of Health and Human Services, 45 CFR Part 164.

13. 49 CFR Part 29, Governmentwide Debarment and Suspension (Nonprocurement), Office of the Secretary, DOT.

CHAPTER 4

Drug Testing in the Nuclear Power Industry: The NRC Fitness-for-Duty Rules

Theodore F. Shults, J.D., M.S.
Chairman, American Association of Medical Review Officers, Research Triangle Park, North Carolina

CONTENTS

4.1 BACKGROUND

The private nuclear power generating industry in the U.S. is highly regulated. The primary federal regulator is the Nuclear Regulatory Commission (NRC). The NRC has a great deal of oversight, inspection responsibilities, safety analysis, and involvement in every aspect of the operation of a nuclear generating facility. The nuclear industry has been highly sensitive to the public safety aspect of its operations and recognizes that in addition to the fundamental engineering issues involved in maintaining the safe operation and security of a reactor and its fuel, a great deal of attention must be paid to the basic human performance issues as well. In this industry careful consideration is given to who has access to the facility, their background, their health, and their behavior — particularly in respect to illegal drug use and alcohol abuse.

The rules that govern drug and alcohol testing in the nuclear industry are called the *Fitness-for-Duty Rules* (FFD) and are found in Title 10 Code of Federal Regulations, Section 26, or ubiquitously known in the industry as 10 CFR 26. The initial fitness-for-duty rule was published in 1989. This rule requires that each nuclear power plant licensee establish a fitness-for-duty program. The NRC crafted these rules at the same time that the HHS Mandatory Guidelines were put together. The NRC adopted the fundamental requirement of using MROs, certified laboratories, and many of the procedural safeguards.

There were, however, a few fundamental differences in philosophy and goals between the HHS Mandatory Guidelines and the NRC fitness-for-duty programs. The NRC had an interest in deterrence of illegal drugs, but the NRC and the industry as a whole also were deeply interested in the concept of overall "fitness" of the employee, the safety of operations, the overall security of the facility and "trustworthiness" of individuals who had access to nuclear plants and facilities. As a result of these considerations, the NRC program that was adopted back in 1988 looked similar to the HHS procedures but contained a number of radically different elements.

First, the NRC gave the individual nuclear electric generating plant (called a licensee) a great deal of flexibility. For example, whereas HHS Mandatory Guidelines restrict drug testing to five categories of drugs, the NRC allows the licensee to test for additional drugs and to test for drugs at lower cutoff levels than specified and required in the HHS Mandatory Guidelines. The NRC also allowed for the on-site testing for specimen validity at the collection site, and allowed the licensee to have an on-site immunoassay screening laboratory for initial screening of urine specimens. The NRC also adopted a modified stand-down provision that a licensee can use. All these deviations from HHS provisions were controversial at the time, but they have worked for the nuclear power industry.

Another fundamental difference is that unlike the DOT/HHS testing model, the NRC does not require the MRO to receive all of the laboratory results from the certified laboratory, or even to review the negative results. (Everything in the nuclear industry is reviewed about ten times — but not by the MRO.) The results typically go to a fitness-for-duty coordinator, who again typically has access to all the testing information, including the MRO records. This is a very different process than seen in DOT testing. On the other hand, the MRO in the nuclear industry often has to play an expanded role with respect to management of the health, qualifications, and prescription drug use of the employees in the program.

In 1995, the NRC published a proposed update to its Fitness-for-Duty Regulations. These proposed rules were open for comment in 1996, were debated, and then went into a state of regulatory limbo. Meanwhile, utilities continued to follow the existing rules and modify and update their programs within the regulatory framework of 10 CFR 26. While the proposed rules were in limbo, a lot happened in other areas of testing. HHS implemented program documents giving laboratories guidelines on how to test for adulterants, and subsequently adopted a mandatory specimen validity rule. DOT has rewritten its drug and alcohol testing rules, adopted procedures for managing insufficient urine volume (shy bladder), and established procedures for testing of adulterated and substituted urine. During this time the nuclear industry was going through a challenging period of reorganization in order to adapt to a new "deregulated" industry of gyrating market conditions.

In 2000, prior to establishing an implementation date for the 1995 "new" rule to go into effect, the NRC held a meeting for the industry to provide any additional clarifications. It was a difficult meeting for all parties. What became clear was that many aspects of what were viewed as enhancements in 1995 looked stale in 2000. In 1995, the regulatory language was still of the old school, as opposed to the new question-and-answer format used by DOT and HHS.

The world changed in September 2001, and the NRC refocused its efforts on counterterrorism and security. In the fitness-for-duty area there was more emphasis and interest in managing fatigue and enhancing background checks. The NRC staff was continuing to work on updating the fitness-for-duty rules, and the industry recognized the benefit of incorporating the DOT and HHS experience with respect to managing specimen validity. The NRC and the nuclear industry also took the opportunity to develop a comprehensive rule (framed as amendments to the existing rules) that better addresses the issues of adulteration, emerging technologies, drug and alcohol abuse, and safety in a less cumbersome way and to integrate a decade and a half of experience from well-run programs.

Thus, after a 10-year period of struggle and at times a mind-numbing regulatory process, the NRC pre-released a proposed new rule in April 2005. This new proposal represents countless hours of work on the part of all of the stakeholders, and a significant catch-up and enhancement of the existing program. True to its original intent, the "fitness-for-duty" rules include the procedures and requirements for the management of "fatigue," which are outlined below.

4.2 THE 2005 PROPOSED AMENDMENTS TO THE NRC FITNESS-FOR-DUTY RULES

On May 16, 2005, the NRC published in the *Federal Register* a notice that stated that it intended to propose:

A rule (that) would amend the Commission's regulations to ensure compatibility with the Department of Health and Human Services guidelines, eliminate or modify unnecessary requirements in some areas, clarify the Commission's original intent of the rule, and improve overall program effectiveness and efficiency and establish threshold for the control of working hours at nuclear power plants to ensure that working hours in excess of the thresholds are controlled through a risk-informed deviation process. Because of the issues raised in response to the earlier affirmed (fitness for duty) rule, a new proposed rule will be published, including provisions to provide significantly greater assurance that worker fatigue does not adversely affect the operational safety of nuclear power plants. This new proposed rule is scheduled to be provided to the Commission by June 1, 2005.*

This new fitness-for-duty rule is available on the NRC Web site.† It is awaiting approval from the Office of Management and Budget, which should take some additional time. The rule when published will provide at least 120 days before going into effect.

As could be anticipated in updating a 10-year-old drug and alcohol testing rule, the proposed fitness-for-duty rule contains a significant number of changes to the existing program. The following are some of the most significant changes that deal directly with drug and alcohol testing. The proposed new *fitness-for-duty* rule would:

- Add requirements for validity tests on urine specimens to determine if a specimen has been adulterated, diluted, or substituted. At the request of stakeholders, the rule would permit licensee testing facilities to perform validity screening tests using non-instrumented testing devices, as proposed by HHS on April 13, 2004 (69 FR 19672), but not yet incorporated into final HHS guidelines.
- Add a requirement that assays used for testing for drugs in addition to those specified in this part, or testing at more stringent cutoff levels than those specified in this part, would be evaluated and certified by an independent forensic toxicologist. (§26.31(d))
- Add a requirement that cutoff levels would be applied equally to all individuals subject to testing. (§26.31(d))
- Establish a process for determining whether there is a medical reason that a donor is unable to provide a urine specimen of at least 30 ml. (§25.119)

4.2.1 Alcohol Testing

With respect to alcohol testing the new provisions would:

---

* The NRC is reviewing the public comments and projects that the rule will become final in January 2007.
† Go to nrc.gov and use the search feature to find "fitness for duty." The searcher should find the old rule, the history of the rule, a wealth of background material, and the new rule in its pre-published form.

- Add requirements for the use of oral fluids (i.e., saliva) as acceptable specimens for initial alcohol tests.
- Lower the blood alcohol concentration (BAC) at which a confirmatory test is required from 0.04 to 0.02%. (§26.31(d)) This is a significant change. One of the reasons for the change has been the observation of the occasional donors who had been on the job for a few hours and still had a breath alcohol test above 0.03%.
- Eliminate blood testing for alcohol. (§26.31(d)) Blood testing is allowed under the existing program as a "voluntary procedure" for individuals who had tested positive on two sequential evidentiary breath testing devices. This extraordinary provision allowing a positive donor to obtain a blood alcohol result was incorporated in an abundance of caution and concern over defending the results in respect to an individual's challenge of the EBT alcohol testing results. The elimination of this provision is recognition that having not just one but two evidential breath testing devices is overkill.

4.2.2   Sanctions

As distinguishable from the general DOT rules (with a few notable exceptions in the FAA rules), the NRC does specify employment sanctions for violations of the fitness-for-duty rule. The sanctions are made more severe under the new provisions. Significantly, they include:

- Require unfavorable termination of authorization for 14 days for a first confirmed positive drug or alcohol test result. (§26.75(e))
- Increase the authorization denial period for a second confirmed positive drug or alcohol test result from 3 to 5 years. (§26.75(e))
- Add permanent denial of authorization for additional FFD violations following any previous denial for 5 years. (§26.75(g))
- Require permanent denial of authorization for refusing to be tested or attempting to subvert the testing process. (§26.75(b))
- Add a 5-year denial of authorization for resignation to avoid removal for an FFD violation. (§26.75(d))

4.2.3   Catch-Up Provisions

The new rule also proposes a number of "catch-up" provisions that incorporate standards already established in DOT and HHS rules. These provisions include such items as:

- Add cutoff levels for initial validity tests of urine specimens at licensee testing facilities and certified laboratories, and require tests for creatinine, pH, and one or more oxidizing adulterants. The rule would not allow licensees and other entities to establish more stringent cutoff levels for validity testing, and would also specify the criteria for determining that a specimen must be forwarded to an HHS-certified laboratory for further testing. (§26.131)
- Replace and amend cutoff levels for initial tests for drugs and drug metabolites to be consistent with HHS cutoff levels. (Decrease the cutoff level for marijuana metabolites from 100 to 50 ng/ml. Increase the cutoff level for opiate metabolites from 300 to 2,000 ng/ml.) (§26.133)

4.2.4   Provisions for MROs

With respect to MROs, the following provisions are being proposed:

- Clarify requirements concerning donor requests to test the specimen in Bottle B of a split sample. (§26.135)
- Clarify and expand the requirements relating to qualifications, relationships, and responsibilities of the MRO.
- Add a requirement that the MRO pass a certification examination within 2 years of rule implementation. (§26.183)

- Add specific prohibitions concerning conflicts of interest. (§26.183)
- Specify MRO programmatic responsibilities. (§26.183)
- Establish the requirements and responsibilities of the MRO staff.
- Add a requirement for the MRO to be directly responsible for the activities of individuals who perform MRO staff duties. (§26.183)
- Add a requirement that MRO staff duties must be independent from any other activity or interest of the licensee or other entity. (§26.183)
- Prohibit the MRO from delegating his or her responsibilities for directing MRO staff activities to any individual or entity other than another MRO. (§26.183)
- Specify the job duties that MRO staff may and may not perform. (§26.183)
- Clarify and expand MRO responsibilities for verifying an FFD violation.
- Make the MRO responsible for assisting the licensee or other entity in determining whether the donor has attempted to subvert the testing process. (§26.185)
- Provide detailed guidance on circumstances in which the MRO may verify a non-negative test result as an FFD policy violation without prior discussion with the donor. (§26.185)
- Clarify MRO responsibilities when the HHS-certified laboratory reports that a specimen is invalid. (§26.185)
- Specify actions the MRO may take if he or she has reason to believe that the donor may have diluted a specimen in a subversion attempt, including confirmatory testing of the specimen at the assay's lowest level of detection for any drugs or drug metabolites. (§26.185)
- Add requirements for the MRO to determine whether a donor has provided an acceptable medical explanation for a specimen that the HHS-certified laboratory reported as adulterated or substituted. (§26.185)
- Incorporate HHS recommendations on verifying a positive drug test for opiates. (§26.185)
- Incorporate federal policy prohibiting acceptance of an assertion of consumption of a hemp food product or coca leaf tea as a legitimate medical explanation for a prohibited substance or metabolite in a specimen. (§26.185)
- Provide detailed requirements for evaluation of whether return-to-duty drug test results indicate subsequent drug use. (§26.185)

4.2.5   Return to Duty

In respect to return-to-duty provisions the new proposed rule would:

- Add a new position, substance abuse expert (SAE), to the minimum requirements for FFD programs and specify the qualifications and responsibilities of the SAE. (§26.187) This SAE may or may not be defined as the SAP is defined in DOT regulations. Nevertheless, the change in the acronym is a significant improvement.
- Specify the role of the SAE in making determinations of fitness and the return-to-duty process, including the initial evaluation, referrals for education and/or treatment, the follow-up evaluation, continuing treatment recommendations, and the follow-up testing plan. The rule would specify the role of the SAE in determinations of fitness based on the types of professional qualifications possessed by the SAE. (§26.189)

4.2.6   Managing Fatigue

The new rule integrates fatigue management and fitness for duty. It is a logical association, but the development of sound fatigue rules that are not burdensome on the industry is a daunting challenge. In summary, the proposed rules:

- Establish program requirements for fatigue management at nuclear power plants.
- Codify a process for workers to self-declare that they are not fit for duty because of fatigue. (§26.197)
- Require training for workers and supervisors on symptoms of and contributors to fatigue and on fatigue countermeasures. (§26.197)

- Require licensees to include fatigue management information in the annual FFD program performance report that would be required under §26.217, including the number of waivers of the individual limits and break requirements that were granted, the collective work hours of any job duty group that exceeded the group average limit in any averaging period, and certain details of fatigue assessments conducted. (§26.197)

4.3 CONCLUSION

This is just an outline of the significant differences. Overall, the NRC has had an exceptional record of performance, even while working with a rule that was clearly long in the tooth. A key to the success of the NRC fitness-for-duty program has been a combination of the rigorous analysis that nuclear managers bring to the table in respect to safety concerns as well as some degree of regulatory flexibility.

One can also anticipate that there may be some changes between the pre-published amendments and what becomes final law, and no doubt the NRC will be evaluating the final rule that HHS is expected to publish with respect to the effectiveness of alternative drug testing technologies. It is also safe to assume that it will be unlikely that the NRC will fall behind again in the race for effective management of substance abuse and employee fitness.

CHAPTER **5**

Workplace Drug Testing outside the U.S.

Anya Pierce, M.Sc., M.B.A.
Toxicology Department, Beaumont Hospital, Dublin, Ireland, U.K.

CONTENTS

5.1 INTRODUCTION

Although workplace drug testing (WDT) began in the U.S., it is increasingly prevalent in all parts of the globe.[1] A major contributor to its spread was U.S. multinational corporations introducing the practice internationally. The amount of WDT performed is still minimal compared to the U.S. but it is steadily increasing.

WDT began in the U.S. military and this trend has continued worldwide, with armed forces everywhere adopting the practice. It is impossible to obtain accurate statistics for most countries, and even when statistics are available, there often is no agreement about what constitutes WDT. For example, many countries include prison testing programs in their statistics.

There is an aura of secrecy about WDT in Europe, as if it were somehow shameful for companies to admit involvement. This is hard to understand because WDT in Europe and Australia is led by health and safety concerns. The greater prevalence of trade unions in Europe appears to inhibit the acceptance of WDT. There is also much greater use of point-of-care testing (POCT) outside the U.S., with its consequent problems of little or no elements of quality assurance.

5.2 SCOPE

This section covers the state of workplace drug testing and the development of regulations, both in place or proposed, accepted methods, and cutoffs in Europe, Australia, and other international arenas.

5.3 WORKPLACE DRUG TESTING IN EUROPE

5.3.1 History

In 1989, Spain held the presidency of the European Economic Community (EEC) and made the following proposals concerning drug testing:

- Examine the criteria currently used for reporting positive results, including the need to distinguish between screening and confirmation results.
- Examine the existing quality assurance programs.
- Examine the validity of certified reference materials for illicit drugs and their metabolites.

This work was followed by a questionnaire to collect information on:

- Substances tested
- Cutoff values
- Test methods
- Need for duplicate samples

- Interpretation and transmission of results
- External quality assurance practices

Last, a survey on quality and reliability was distributed. This led to the formation of an expert group under the aegis of Marie Therese Van der Venne of DG VI of the European Commission. Several meetings were held in Luxembourg with one or two representatives from each EEC country. This culminated in a 2-day meeting in Barcelona with original representatives and additional experts. Recommendations[2] were finalized and published in journals nominated by the experts from each country under the following categories:

- Sample handling and chain of custody
- Cutoff values
- Analytical methodology
- Educational requirements
- External quality assurance and accreditation

These recommendations met with a great deal of opposition. Some thought they were too limited, while others thought they were too restrictive. Some disliked the specific cutoffs selected. Different countries had widely varying attitudes, socially and legally, to drug use.

5.3.2 European Workplace Drug Testing Society

The sponsor in the European Commission died suddenly and workplace testing guidelines failed to progress until a meeting entitled "Drug Testing at the Workplace" was held in Huddinge Hospital in Stockholm, Sweden. This meeting led to the formation of the European Workplace Drug Testing Group (EWDTG) on March 1, 1998. The group consisted of one or two members from each EU country and representatives from Norway and Switzerland. From this core group the European Workplace Drug Testing Society (EWDTS) developed.

The mission of the EWDTS is to ensure that workplace drug testing in Europe is performed to a defined quality standard, in a legally secure manner, and to provide an independent forum for discussion of all aspects of workplace drug testing. Guidelines for urine testing are published and are discussed below. Guidelines for other matrices are under development.

The EWDTS objectives are to:

- Provide the source of expertise on WDT in Europe
- Function as the primary advisor to the European Commission
- Develop relevant literature and dispense information via the EDWTS Web site
- Train medical review officers (MROs)
- Train sample collectors

There is no specific legislation regarding WDT in any country in Europe. Finland had planned to introduce it, but at the last minute deleted the section from the relevant act.

Industries that perform WDT are mainly in:

- Transport
- Information technology
- Petrochemicals
- Shipping
- Pharmaceuticals
- Customer support (call) centers

Testing occurs predominantly at the pre-employment level, although in transport it is also routinely performed after an accident. Government employees outside the U.S. are not tested.

5.3.3   Sweden

In 1995, the Swedish government evaluated WDT and found no need for legislation. Instead, it was found preferable for the labor market to regulate this issue itself. In 1998, 23,997 people were tested, of whom 2.3% were positive.

A person working as a cleaner in a non-safety critical area of a nuclear power plant objected to drug testing. The case was referred to the European Court of Human Rights. The court handed down a judgment on March 9, 2004 rejecting the application, saying it was impractical to differentiate employees and that drug testing of all its employees was a proportionate measure that did not violate Article 8.2 of the European Convention on Human Rights. This is the only case law in Europe at present.

5.3.4   The Netherlands

Interestingly, the only laws referring to WDT in the EU are in the Netherlands, where pre-employment testing is prohibited. There is also opposition from trade unions and occupational health specialists, who believe that it is an infringement of individual privacy. In companies that do drug testing, it cannot be obligatory and employees have the right to refuse.

Only one laboratory performs WDT. It is located in Rotterdam with a client-base of petrochemical and shipping industries, many of whose employees work offshore. About 20,000 tests are performed annually. Amsterdam, renowned for its coffee shops where small amounts of cannabis can be purchased, has almost no WDT.

5.3.5   Spain and Portugal

In a 1994 survey of companies in the Lisbon district with more than 50 workers, Vitòria[3] estimated that 20% were doing drug testing. Pinheiro et al.[4] found that 14% of the largest Portuguese companies (with a workforce greater than 1000) were performing tests in 1997, the majority without correct chain of custody or confirmation.

No statistics are available from Spain. The majority of WDT on the Iberian Peninsula is done by the military. The most representative (and probably sole) indicators of WDT activity are those of the Portuguese army, where positive results have decreased from 17% in 1986 to 5.8% in 1995. Positive cannabis and opiate tests accounted for 4.1 and 1.4% of total positive tests.

5.3.6   Luxembourg

Some private companies are performing on-site testing for drugs of abuse, and some companies are sending urine specimens to special laboratories. Again, there are no firm statistics.

5.3.7   France

Air France and the automobile industry do the majority of drug testing. Private laboratories provide these services. In August 2004 the Department of Transport allowed occupational physicians in the railroad industry to drug test workers in safety-critical positions.

5.3.8   Germany

There is very little WDT in Germany. Most of the major companies in transport and manufacture do not test. Generally, WDT is perceived as an invasion of privacy, although some companies, mainly around Frankfurt, have begun testing.

5.3.9   Greece

WDT is done only in Thessaloniki, at the Aristotelian University.[5] Specimens for analysis come from:

- District attorney offices for testing of prisoners' specimens
- Directorate of Transportation for drivers seeking the reinstatement of driving licenses revoked for previous drug abuse
- Private individuals during pre-employment testing
- Security services under a law passed in 1997
- Prostitutes and housekeepers at houses of prostitution under a law passed in November 1999

5.3.10   Ireland

Ireland has a large number of nonindigenous pharmaceutical and information technology companies. The majority of them test at the pre-employment level and use the policies of their parent companies. The armed forces began random testing in September 2002, although they have performed pre-employment testing for many years. Some companies use laboratories in other countries. The use of POCT is widespread; it is estimated that 50,000 tests are performed annually.

In 2006 the Health and Safety Authority enacted into law the Safety, Health and Welfare at Work Act 2005. One of the more controversial sections states as follows:

An employee shall, while at work —

(a)   comply with the relevant statutory provisions, etc.
(b)   ensure that he or she is not under the influence of an intoxicant to the extent that he or she is in such a state as to endanger his or her own safety, health or welfare at work or that of any other person.
(c)   if reasonably required by his or her employer, submit to any appropriate reasonable and proportionate tests, by or under the supervision of a registered medical practitioner who is a competent person, as may be prescribed. This is the first such law in Europe. It still has to be fine tuned to decide which occupations are considered safety critical and what drugs are to be looked for.

5.3.11   U.K.

Testing became more prevalent in the 1980s and is growing. The discovery of oil and the opening of oil rigs in the North Sea and chemical and transport industries have increased testing demand. At present, it is estimated that more than 500,000 tests are performed annually. It is not known how many of the analyses are generated from outside the U.K.

The Railway & Transport Safety Act covers air/road/rail and sea, which includes any person in charge of a vehicle, and means they can be tested. The rail sector is the most rigorous with pre-employment "with cause" testing for all and random testing for those in safety-critical positions, and education and help for anyone who discloses a problem. Overall, there is a trend of taking a risk assessment approach to drugs and alcohol, and looking at business-critical as well as safety-critical issues.

5.3.12   Belgium

The main users of WDT are in transport and the automobile industry. There are again no reliable statistics. The General Medical Council established strict guidelines in 1993: analysis may only be performed in a laboratory, drug testing is only allowed if clinical examination of impairment is not possible, and positive results must be confirmed by another laboratory. A recent (2002) initiative to amend these rules was not successful.

5.3.13   Finland

An Act on the Protection of Privacy in Working Life, which was recently enacted into law, covers several aspects of workplace drug testing. The Finnish guidelines on drug testing are the European Workplace Drug Testing (EWDTS) guidelines. WDT in Finland is mostly pre-employment testing, but random testing is increasing. WDT is usually part of health screening and the employer may often be told only that the applicant has failed the medical examination, with no further details provided.

5.3.14   Denmark

There is no legislation regarding WDT in Denmark. It is organized at the local level and mainly for the transport sector, offshore workers, and the police force. Persistent findings indicate that up to 12% of the workforce is influenced by one or more drugs.

Approximately 10,000 tests per year are performed, mainly random testing or cluster testing, with very few pre-employment tests. Approximately 90% of all drug tests are performed as on-site testing (immunoassay screening) by health professionals, who also play a role in developing drug-free workplace programs.

5.4   ATTITUDE SURVEYS

5.4.1   U.K. CIPD Survey

A 2001 survey of 281 organizations by the Chartered Institute of Personnel and Development reported[6] that:

    20% of the organizations were considered safety-critical, 38% partly so
    56% have an alcohol and drug policy
    18% carry out drug testing
    12% perform drug testing when drug use is suspected
    9% test for pre-employment
    7% test for post-accident evaluation
    6% perform random testing
    6% do post-rehabilitation testing
    2% test prior to promotion

5.4.2   Drug Testing in Prisons

Drug testing in prisons varies within and between countries. Positive tests tend to result in loss of privileges only. A split sample is taken, and re-screening and confirmation are available in some countries upon receipt of payment from the prisoner. The U.K. has a mix of mandatory and voluntary testing. Sweden also tests for anabolic steroids.

5.4.3   Employee Attitudes Survey

A survey was conducted in Sweden, Portugal, and Ireland in 2000 to acquire knowledge about attitudes toward drug testing in the workplace.[7] Questionnaires were given to people who were tested for drugs at pre-employment. The answers were voluntary and anonymous. The questions asked were:

Do you think that drug testing can be a good method to achieve a more drug-free workplace?

Do you consider donating a urine sample for this purpose in any way offensive?

If you had a choice, what matrix would you prefer?

Do you think that narcotics are a problem in our society?

What are your views on the use of recreational drugs such as cannabis or ecstasy?

The results showed little difference between countries, with the Swedish most certain of the benefits of WDT and the Portuguese the least. The Irish preferred urine as a matrix with very few countries in favor of sweat testing. Individuals under 20 and those over 40 saw the least danger in the use of recreational drugs. The main problem with the survey was that the number of participants was small and self-selecting, and the survey should have involved more countries. Interestingly, a number of people who responded had reservations about the failure to differentiate between cannabis and ecstasy use. They felt that ecstasy was much more dangerous and should have been treated separately.

5.5 EUROPEAN LABORATORY GUIDELINES FOR LEGALLY DEFENSIBLE WORKPLACE DRUG TESTING

5.5.1 Introduction

The EWDTS guidelines[8] are based on the U.K. WDT guidelines. Urine is the only matrix included at present, although it is planned to introduce breath, oral fluid, hair, etc. The guidelines represent best practice, which can withstand legal scrutiny, and are intended to provide a common standard throughout Europe. The guidelines have now been accepted by the European Accreditation (EA) body as the benchmark for WDT.

The guidelines include:

• Specimen collection
• Laboratory organization
• Laboratory analysis
• Quality control and quality assurance
• Medical review officer

5.5.2 Screening Cutoffs

Table 5.1 presents European screening cutoffs.

Table 5.1   European Screening Cutoffs

Drug	ng/ml
Amphetamine	300
Benzodiazepines	200
Cannabis metabolites	50
Cocaine metabolites	300
Opiates (total)	300
Methadone/metabolites	300
Barbiturates	200
Phencyclidine	25
Buprenorphine/metabolites	5
LSD/metabolites	1
Propoxyphene/metabolites	300
Methaqualone	300

Table 5.2   European Confirmation Cutoffs

Amphetamine	200
Methylamphetamine	200
MDA	200
MDMA	200
MDEA	200
Other amphetamines	200
Temazepam	100
Oxazepam	100
Desmethyldiazepam	100
Other benzodiazepines	TBA
11-Nor-9-tetra hydrocannabinol-9-carboxylic acid	15
Benzoylecgonine	150
Morphine	300
Codeine	300
Dihydrocodeine	300
6-Monoacetylmorphine	10
Methadone/metabolites	250
Barbiturates	150
Phencyclidine	25
Buprenorphine/metabolite	5
LSD/metabolites	1
Propoxyphene/metabolite	300
Methaqualone	300

5.5.3   Confirmation Cutoffs

Table 5.2 presents European confirmation cutoffs.

5.6 U.S.–EU COMPARISON

Neither the individual European governments nor the EU parliament has shown real interest in WDT compared to the U.S. government. The U.S. has legally enforceable guidelines, while the European guidelines have no legal standing. In Europe, there is greater use of POCT and small non-accredited laboratories. In the U.S., testing is performed in a small number of large laboratories that are accredited and operated under very strict controls. Testing in the U.S. is mandatory for federal employees, which is not the case in Europe.

5.7 WDT IN AUSTRALIA AND NEW ZEALAND

5.7.1   Standards

WDT is increasing in Australia. Although mining is the main industry to embrace WDT, other industries are also testing, including:

- Metalliferous
- Quarrying
- Transport
- Construction

Neither transport nor construction industries have introduced a systematic procedure to measure the extent of drug use within their workforce. Some transport industries across Australia do have

Table 5.3   Australian/New Zealand (AS/NZS 4308) Screening Cutoffs

Drug	ng/ml
Opiates	300
Sympathomimetic amines	300
Cannabis	50
Cocaine	300
Benzodiazepines	200

Table 5.4   Australian/New Zealand (AS/NZS 4308) Confirmation Cutoffs

Drug	ng/ml
Morphine, codeine	300
Amphetamine, methamphetamine, MDMA	300
Phentermine, ephedrine, pseudoephedrine	500
Carboxytetrahydrocannabinol	15
Benzoylecgonine, ecognine methyl ester	150
Oxazepam, temazepam, diazepam, nordiazepam, 7-aminoclonazepam, 7-aminonitrazepam, 7-aminoflunitrazepam	150

ad hoc testing. Similar to Europe, drug testing in Australia is conducted as part of an overall Occupational Health and Safety policy. Thus, there are many safeguards for employees who test positive. They are not dismissed, but instead are referred to a rehabilitation program. Only after repeatedly testing positive are individuals liable for dismissal.

Many unions and even Employee Assistance Providers do not support drug testing. In general, within the workforce, the most common drugs of abuse identified are cannabis and methamphetamine. The most common drugs detected are therapeutic codeine and pseudoephedrine. Tables 5.3 and 5.4 present Australia/New Zealand screening and confirmatory cutoffs.

5.7.2   The Australian Mining Industry

This industry is the leader in the introduction of drug and alcohol testing of its workforce. Drug testing is carried out as part of occupational health and safety requirements.

The mining industries in Western Australia, Queensland, and New South Wales are in the forefront. Policies, including compulsory drug and alcohol testing, usually require that an employee is not impaired due to the ingestion of illicit drugs. This wording has created problems, as it is impossible to prove impairment, particularly in the case of a positive cannabis test. There is a lobby to increase the cannabis cutoff to reduce this perceived problem. Unfortunately, policy makers have used terminology from established alcohol testing for drug testing. Even though drug testing has been introduced under health and safety regulations, the unions have concerns. They defend an individual's right to privacy, are concerned about the accuracy of drug testing, and state that individuals should not be penalized for using cannabis on their own time.

Each Australian state has its own mining legislation. In effect, it is the duty of mine managers to ensure that no person is impaired by alcohol or drugs while on site. On-site drug testing is well established. Accredited laboratories perform confirmation analyses. Although there are no legal requirements, most companies select laboratories that are accredited by the National Association of Testing Authorities (NATA) to conduct analyses according to Australian/New Zealand Standards AS/NZS 4308.

More recently, drug and alcohol testing has been extended to include contractors to the mining industry. To date, contractors present a greater risk for impairment than do employees of a company. In due course, it is expected that testing will significantly reduce the incidence of recreational drugs by workers, either employed full time or contracted to the mining industry.

5.7.3   New Zealand

Laboratories perform WDT work to the standards of the AS/NZS 4308.[9] The 2001 standard is currently being updated to include a section for use of "on-site" screening devices. Standards Australia is also developing an Oral Fluid Standard.

The number of tests is rising each year. In 1998/1999, about 3000 urine specimens were tested. This rose to an estimated 50,000 in 2005/2006.

The industries that test are divided as follows (2005/2006):

Road/Horizontal Construction	21%
Forestry	15%
Transport	13%
Meat/Poultry	11%
Dairy	10%
Fishing/Shipping	6%
Aluminum Smelting	4%
Power/Oil	3%
Manufacturing	2%
Vertical Construction	2%
Mining	<2%
Personnel Consulting	<2%

The majority of testing (78%) is at pre-employment, followed by random (15%). Only 2% of the tests are performed after an accident/incident or for "just cause."

Percentage positives in 2005/2006 were 8% (pre-employment), 13% (random testing), and 30% (post-accident or just cause).

5.8 WDT IN SOUTH AMERICA

5.8.1   Introduction

Workplace drug testing is not a common practice in South America. Most countries do not have established guidelines or policies. Generally, it is performed because the company (usually North American based) requires it as part of a global policy of WDT. The most notable case is Exxon, which requires all of its employees to undergo testing. The tests are mainly performed in a U.S. laboratory and conform to SAMHSA/NIDA guidelines.

5.8.2   Brazil

WDT began in 1992 with oil companies and has broadened to include transport and car manufacture, as well as the state police. Since 1992, more than 40,000 tests have been performed and they conform to SAMSHA/NIDA requirements. Over 80% of these tests have been done in the past 3 years. Tests are mainly in urine, but can include hair and oral fluid. Positive tests are handled according to the individual company's drug prevention program. This may include outright dismissal, but the vast majority of companies adopt a more tolerant approach with rehabilitation and treatment. An individual is usually allowed up to three positive drug tests prior to dismissal.

This latter approach has yielded positive results, with many employees voluntarily seeking help for their drug/alcohol problems. When employees perceive that they will receive adequate and sympathetic treatment, rather than dismissal, they are more prone to accept rehabilitation.

The rate of companies adopting a drug prevention policy with testing is increasing at the rate of 5 to 7% a year. Government policy is neither for nor against WDT. It is seen as one important step in the antidrug campaign: it promotes issues of occupational health and safety.

5.8.3 Other South American Countries

Argentina: There is no national program and only one company (Exxon) requires it.

Bolivia: There is no official drug-testing program apart from companies that require testing as part of a global program.

Chile: It is starting to develop policies but without any testing so far.

Colombia: Again only one oil company (Exxon) has a WDT program.

Paraguay: A WDT program is beginning, but the tests will be performed in Brazil.

Venezuela: Some major oil companies have WDT programs with testing performed abroad. According to Dr. Lilia Scott (M.D.), there is great difficulty in implementing such programs in the country, as the government does not endorse WDT.

Ecuador, Peru, Uruguay, Guyanas (Guyana, Suriname, and French Guyana): There are no reports of WDT in these countries.

5.9 CONCLUSION

WDT is minimally performed around the world compared to in the U.S. It is not envisaged that it ever will become as prevalent. In the U.S., there are rigid regulations for all aspects of drug testing from collection to reporting. The rest of the world does not have the same safeguards. There is much more use of POCT with all the attendant problems of correct use of the device. Some companies do not inform the employee or presumptive employee that he or she is being tested. Often there is no confirmation test. The use of alternative matrices for WDT is also increasing.

ACKNOWLEDGMENTS

Dr. John Lewis, Toxicology Unit, Pacific Laboratory Medical Services, Sydney, NSW, Australia

Prof. Alain Verstraete, Klinische Biologie, University of Ghent, Belgium

Dr. Anthony Wong, Maxilab Diagnósticos, Brazil

Dr. Leendert Mostert, Deltalab, Rotterdam, the Netherlands

Mr. Per Bjorklov, Huddinge Hospital, Stockholm, Sweden

Prof. Robert Wennig, Centre Universitaire, Luxembourg

Ms. Satu Suoimen, Helsinki, Finland

Prof. Gerold Kauert, Centre of Legal Medicine, Frankfurt, Germany

Dr. Cesar Fernandez, Toxscreen, Barcelona, Spain

Ms. Lindsay Hadfield, Medscreen, London, U.K.

Mr. Leo Rebsdorf, ScanScreen ApS, Kolding, Denmark

REFERENCES

1. Verstraete, A.G. and Pierce, A., Workplace drug testing in Europe, *Forensic Sci. Int.*, 121(1–2), 2–6, 2001.
2. de la Torre, R., Segura, J., de Zeeuw, R., Williams, J., et al., Recommendations for the reliable detection of illicit drugs in urine in the European Union, with special attention to the workplace, *Ann. Clin. Biochem.*, 34, 1997.
3. Vitoria, P.D., Consumo de alcool e drogas ilegais em empresas do distrito de Lisboa. Fundacao Portuguesa para o Estudo, Prevencao e Tratamento da Toxicodependencia, Cascais, 1994.

4. Pinheiro, J., Pinheiro, R., Marques, E.P., and Vieira, D.N., O consumo de substãncias nas maiores empresas portuguesas. 4° Forum de Medicina do Trabalho, Lisboa, 1997.
5. Tsoukali, H., Raikos, N., Kovatsi, L., Kotrotsi, E., Athanasaki, R., and Psaroulis, D., Workplace drug testing in Northern Greece during the period 2000–2002, presented at 3rd European Symposium on Workplace Drug Testing, Barcelona, Spain, 2003.
6. Survey on alcohol and drug policies in U.K. organisations, Chartered Institute of Personnel and Development, www.cipd.co.uk.
7. Pinheiro, J., Pierce, A., and Bjorklov, P., Employee Attitude Survey, presented at 2nd EWTDS Symposium, Rimini, Italy, 2000.
8. European Laboratory Guidelines for legally defensible Workplace Drug Testing, www.ewdts.org.
9. Nolan, S. and Turner, S., Workplace & Prison Inmate Drug Testing Programmes in New Zealand over a ten-year period, presented at TIAFT 41st International Meeting, Melbourne, Australia, 2003.

CHAPTER 6

Analytical Considerations and Approaches for Drugs

Daniel S. Isenschmid, Ph.D.[1] and Bruce A. Goldberger, Ph.D.[2]
[1] Toxicology Laboratory, Wayne County Medical Examiner's Office, Detroit, Michigan
[2] University of Florida College of Medicine, Gainesville, Florida

CONTENTS

Although recent national surveys have reported declines in overall drug use, more than 11 million Americans still reported using one or more illicit drugs in 1992.[1] In addition, results from several studies targeting specific populations, particularly students in high school and college, show that this downward trend has started to reverse itself.[2,3] In response to the numerous reports on the incidence of alcohol and drug abuse in the workplace, many employers have introduced substance abuse programs, one component of which is urine drug testing.[1,4,5] There are several very important differences between this type of testing and the laboratory testing performed for medical reasons:

- The test is not performed on a patient; instead, a specimen is collected from a donor.
- The test result is not used to support a diagnosis; it is a single collection that is used in decisions relating to hiring, suspension from employment, referral to employee assistance programs (EAP), and, occasionally, dismissal of the employee. Unlike a true clinical test, it has no medical data to support its veracity and, except in rare circumstances, the collection cannot be repeated.
- For regulated industries in which drug testing is required, the laboratory has to be certified by the Department of Health and Human Services.[6,7] This certification from the Substance Abuse and Mental Health Services Administration (SAMHSA, formerly the National Institute on Drug Abuse) has become the "standard of care" for all drug testing, including that performed for the nonregulated industry. This certification mandates that the laboratory use certain analytical procedures, including gas chromatography mass spectrometry (GC/MS), for confirming immunoassay results, and that it follows rigid chain-of-custody, quality assurance, and data review guidelines.
- The result may be used in a legal environment to support disciplinary action.

Workplace urine drug testing is performed by many laboratories. Because of the strict chain-of-custody and security requirements, most drug testing laboratories are "stand-alone" laboratories, either secured separately from other specialties in larger laboratories or as distinct entities. More recently, there has been a trend toward "mega-laboratories." These are single locations dedicated to workplace drug testing, processing thousands of specimens per shift. Both service expectations and the economics of drug testing have caused the evolution of such facilities. Today's expectations are that negative results will be available within 24 h of collection of the specimen, and that positives will be available within another 24 h. Obviously these expectations, together with the desire to decrease costs, will play an increasing part in the choice of the analytical procedures used. Increasing emphasis is being placed on the ease of automation of such procedures.

This chapter outlines the analytical protocols used in such programs. Since there may be significant penalties if a person tests positive for a drug of abuse, it is critical that individuals are not falsely accused of drug use. It is, therefore, important that drug tests are precise and accurate, utilizing validated methodologies. The validity of a drug test is defined as the ability of an assay to detect drug and/or its metabolites in biological fluids following drug administration. This definition encompasses a variety of factors including: assay characteristics such as sensitivity and specificity; and metabolic and pharmacological variables such as dose, route of administration, biological fluid pH, and intersubject variability in absorption, metabolism, and excretion.[8] Typical urine drug and drug metabolite detection times are given in Table 6.1.

Table 6.1   Typical Urine Drug and Drug Metabolite Detection Times

Drug Class	Detection Times[a]
Amphetamines	1–3 days
Barbiturates	Short-acting, 1 day; intermediate- and long-acting, 1–3 weeks
Benzodiazepines	5–7 days
Cannabinoids	1–3 days; greater (several weeks) with chronic use
Cocaine	1–3 days
LSD	Less than 1 day
Methadone	1–3 days
Methaqualone	1–2 weeks
Opiates	1–3 days
Phencyclidine	Up to 3 days
Propoxyphene	1–2 days

[a] Detection times vary with method of detection, dose, route of administration, frequency of use, and individual factors.

Sources: Substance-Abuse Testing Committee, 1988; Baselt and Cravey, 1995.

6.1 IMMUNOASSAY TESTING

When testing urine specimens for drugs of abuse, laboratories usually screen specimens utilizing a nonspecific immunoassay. These assays may detect classes of drugs, such as the opiates, rather than specific drugs within a class, such as morphine. Since the majority of specimens will test negative, this eliminates the need to do more costly additional testing.

Immunoassays are based on the principle of competition between labeled and unlabeled antigen (drug) for binding sites on a specific antibody. The label may be an enzyme (homogeneous enzyme immunoassays or heterogeneous enzyme-linked immunosorbent assays), a radioisotope (radioimmunoassay), a fluorophore (fluorescence polarization immunoassay), or latex microparticles (particle immunoassay). Immunoassays are characterized by several variables including specificity and sensitivity.

Specificity is the ability of an assay to distinguish the target analyte(s) from other compounds including those with and without structural similarity. Specificity data are typically provided by manufacturers of immunoassays in the form of a package insert. The specificity requirements and design of a particular assay are critical, and may be dependent on the metabolism of the specific drug. For example, since cocaine is rapidly metabolized, all cocaine immunoassays are formulated to detect the primary cocaine metabolite in urine (benzoylecgonine). Furthermore, assays are often designed to be drug class specific (rather than analyte specific) in order that several analytes within a drug class, such as opiates, may be detected with a single assay. Immunoassay performance may be affected by "adulterants," which may be added to a urine specimen in an effort to subvert the testing process.

Very early in the evolution of workplace drug testing, it was decided, after considerable debate, that there was a need for assay "cutoffs" to ensure some standardization and equality among the various programs. The "cutoff" is an administrative breakpoint used to distinguish positive and negative specimens. When utilizing assay cutoff concentrations, any specimen that contains drug at a concentration at or above the cutoff will be reported as positive, and any specimen that is less than the cutoff will be reported as negative. Immunoassay screening cutoff concentrations are summarized in Table 6.2.

When selecting an immunoassay, there are several factors to consider. These include (1) specificity of the kit to the drug group or drug being tested for, (2) the stability of the immunoassay over time, (3) the ability of the immunoassay either to detect adulterants or to be resistant to them, and (4) the ability of the immunoassay to be automated.

Table 6.2   Immunoassay Cutoff Concentrations (ng/ml)

Drug Group	Target Antigen	Cutoff
Amphetamines	D-Amphetamines, D-methamphetamine, or both	300 or 1000*
Barbiturates	Secobarbital	200 or 300
Benzodiazepines	Oxazepam or nordiazepam	100, 200, or 300
Cannabinoids	THCA	20, 25, 50*, or 100
Cocaine	Benzoylecgonine	150 or 300*
Methadone	D-Methadone or D,L-methadone	300
Methaqualone	Methaqualone	300
Opiates	Morphine	300 or 2000*
Phencyclidine	Phencyclidine	25* or 75
Propoxyphene	D-Propoxyphene	300

* Cutoffs used in the SAMHSA Guidelines and the regulated programs.

6.1.1   Radioimmunoassay

The first laboratory assays developed for the detection of drugs of abuse and their metabolites in urine were based on the radioimmunoassay technique. Heterogeneous radioimmunoassays are based upon the competitive binding of ^{125}I-radiolabeled antigen and unlabeled antigen for antibody. Analyte antibody and radiolabeled antigen are added to the specimen and the analyte present in the specimen competes with radiolabeled antigen for antibody sites. After precipitation of the antigen–antibody complex with a second antibody reagent, the supernatant is removed and the pellet containing bound antigen is counted using a gamma counter. The amount of radioactivity is inversely related to the concentration of analyte in the specimen.

Radioimmunoassays are very sensitive, generally more so than non-isotopic enzyme immunoassays. Radioimmunoassays have been developed for all common drugs of abuse and have been designed to produce qualitative and semiquantitative results. The disadvantages of radioimmunoassays include limited reagent shelf-life due to the short half-life of ^{125}I, the requirement for additional laboratory safety precautions due to radioisotope use, and considerable technician involvement. They are infrequently used in the urine drug testing industry today, although they are still used for drug testing in alternative matrices, especially blood and hair, as they are less susceptible to matrix effects than many enzyme immunoassays.

6.1.2   Enzyme Immunoassay

Enzyme immunoassays for the detection of drugs of abuse in urine initially developed during the early 1970s by Syva Company under the trade name EMIT™ (enzyme multiplied immunoassay technique) are now also available from a variety of manufacturers. At the time of its introduction, the EMIT technology offered several advantages over other available procedures for drug testing. The technique provided moderate sensitivity, eliminated a separation stage, utilized reagents with longer shelf-life, could be performed rapidly, and was easily adaptable to manual and automated spectrophotometer-based laboratory instrumentation. The principle of the EMIT assay is based on competition between drug in the specimen and drug labeled with the enzyme glucose-6-phosphate dehydrogenase for antibody binding sites. The enzyme activity decreases following binding to the antibody; therefore, the drug concentration in the specimen can be measured in terms of enzyme activity. In practice, when free analyte is present in the specimen, the displaced free enzyme will convert nicotinamide adenine dinucleotide (NAD) to NADH resulting in a change in absorbance, which is measured spectrophotometrically. The resultant increase in absorbance is proportional to the concentration of the analyte in the specimen.

More recently, this technology has been applied to enzyme-linked immunosorbent assays (ELISA). These assays differ from liquid enzyme immunoassays in that the assays are heterogeneous and are performed on a microtiter plate and are therefore less subject to the matrix effects than are homogeneous enzyme immunoassays. In ELISA diluted specimens and drug-labeled enzyme are incubated in microplate wells coated with fixed amounts of oriented polyclonal antibody. The enzyme (horseradish peroxidase)-drug complex competes for antibodies with the free antigen (drug). After incubation, the wells are washed and a chromogenic substrate (tetramethylbenzidine) is added. Tetramethylbenzidine is converted to a colored product by the unbound drug-enzyme complex. The color reaction is stopped by adding dilute acid. Plates are then read at 450 nm with the intensity of color inversely proportional to the concentration of drug in the sample. These assays offer all of the advantages of radioimmunoassays without radioactive disposal issues and in addition are more amenable to automation using robotic pipettor-diluters, plate washers, and spectrophotometric readers. In addition, as the assay uses an enzyme as a label, shelf-lives are much longer than for radioimmunoassays. Although enzyme-linked immunoassays are not commonly used in very high-volume urine drug testing laboratories, they are used in high-volume oral fluid drug testing laboratories and for blood drug testing. Many manufacturers now offer these types of assays.

In the mid-1990s, Microgenics Corporation introduced a new homogeneous assay system for drugs of abuse in urine called cloned enzyme donor immunoassay, or CEDIA®. The CEDIA assay is based on the genetically engineered bacterial enzyme β-galactosidase that consists of two inactive fragments. These fragments spontaneously associate to form an active enzyme that cleaves a substrate, chlorophenol red β-galactopyranoside, producing a color change that is measured spectrophotometrically. In the assay, analyte in the specimen competes with analyte conjugated to one inactive fragment of β-galactosidase for antibody binding site. If analyte is present in the specimen, it will bind to antibody, leaving the inactive enzyme fragments free to form active enzyme. The amount of active enzyme formed and resultant increase in absorbance are proportional to the amount of analyte present in the specimen.

6.1.3 Fluorescence Polarization Immunoassay

Fluorescence polarization immunoassays (FPIA) were initially developed by Abbott Laboratories for use by laboratories performing therapeutic drug monitoring and were subsequently adapted for detection of drugs of abuse in urine specimens. The Abbott FPIA system employs a fluorescein-analyte labeled tracer. The tracer, when excited by linearly polarized light, emits fluorescence with a degree of polarization inversely related to its rate of rotation. The analyte of interest, if present in the specimen, competes for a limited number of antibody binding sites with the labeled tracer. If the tracer molecules become bound to its specific antibody, the rotation of the tracer assumes that of the larger antibody molecule, which is significantly less than the unbound tracer molecule, and the subsequent polarization is high. The remaining unbound tracer becomes randomly oriented and rotates rapidly; the subsequent polarization is low. The degree of fluorescence polarization is inversely related to the concentration of analyte in the specimen.

The Abbott FPIA abused drug assays offer laboratories several options that are not available with other systems. All FPIAs utilize a six-point calibration curve, in conjunction with nonlinear regression data manipulation, to estimate unknown specimen drug concentrations. Assay cutoff concentrations can be programmed by the laboratory between the assay sensitivity limit and the highest calibrator concentration. Also, the data printout can be programmed to include either qualitative or semiquantitative results. The assays are amenable to the analysis of other matrices, are very stable, and are resistant to a number of adulterants.

6.1.4 Particle Immunoassay

In the early 1990s, Roche Diagnostic Systems introduced the Abuscreen ONLINE® (immunoassay system known as the kinetic interaction of microparticles in solution, or KIMS®). The assay is based on the competition of analyte in a specimen for free antibody with analyte-microparticle conjugate. Particle aggregation occurs in the absence of analyte when free antibody binds to the analyte-microparticle conjugate. The formation of an aggregate scatters light, which results in a reduction of light transmission that can be measured spectrophotometrically. The absorbance change is inversely related to the analyte concentration. The assays can be run on many automated clinical chemistry analyzers, have good stability, and are resistant to a number of adulterants.

6.1.5   On-Site Drug Testing

A new market of drug testing devices has grown in recent years due to the desire to decrease turnaround time and cut costs. These "on-site" drug testing kits are typically used at the specimen collection site and are self-contained devices utilizing immunoassay technology for drug testing. The use of these devices has increased in recent years due to their simplicity, ease of performance, and no requirement for expensive equipment or skilled personnel. Minimal operator training is required, results are typically obtained within 10 to 15 min of specimen application, and the presence or absence of a color is the most common method of reading the results. The cost saving occurs in that only those specimens that test positive by this screening assay will be sent to a laboratory for confirmation by an alternative analytical technique. Depending on the specific device and the number of drugs assayed, the cost per test may range between $3.00 and $25.00.[9]

The kits may be used in emergency rooms, physician offices, drug treatment centers, probation services, and by employers for pre-employment and random or reasonable cause drug testing to provide preliminary qualitative results. The majority of the devices have built-in validation controls. If the testing is performed for forensic purposes, all positive specimens must be confirmed by an alternative technique. These tests utilize urine as the drug testing matrix and are available to test several drug classes. There are many manufacturers of these devices and their efficacy, as with all assays, depends on their validity in the detection of drugs of abuse. There are several reports of validity assessment studies of these devices.[9–20]

6.1.6   Testing for HHS-Regulated Drugs by Immunoassay

6.1.6.1   Amphetamines

Amphetamines are sympathomimetic phenethylamine derivatives with potent central nervous system stimulant activity. Amphetamine and methamphetamine are extensively metabolized and can be detected in urine specimens for up to 3 days. Because methamphetamine is metabolized to amphetamine, amphetamine and methamphetamine-amine are present in urine after methamphetamine ingestion.

In workplace testing, the detection of amphetamines by immunoassay is commonly performed using a kit designed to detect the D-isomers of amphetamine and/or methamphetamine, and not kits designed to detect the amphetamine group. However, there are amphetamine class assays available for clinical toxicology laboratories and others interested in detecting relatively low concentrations of other sympathomimetic amines. Many of these agents, including ephedrine, pseudoephedrine, phentermine, phenylpropanolamine, and phenylephrine, are found in popular nonprescription over-the-counter, cold, allergy, and diet medications. Even though the kits designed to detect the D-isomers of amphetamine and/or methamphetamine have low cross-reactivity to these other sympathomimetic amines, their concentrations in urine may still be high enough to trigger a false-positive test result.

There are important differences between the kits most widely used for amphetamine detection. They all share a 1000-ng/ml cutoff, but some use D-methamphetamine as a calibrator, whereas others use D-amphetamine. When comparing tests for amphetamines, three important questions must be answered: (1) are all assays equivalent in detecting the abuse or use of D-methamphetamine or D-amphetamine, (2) are all assays stereospecific to the D-isomers or will they detect use of L-isomers (Vicks Inhaler) or drugs metabolized to L-isomers (Selegiline), and (3) if the illicit form of methamphetamine used is a racemic mixture, can all assays detect the use? As one might expect, the answers to these questions vary significantly.

The EMIT and CEDIA enzyme assays have good cross-reactivity toward D-amphetamine (the metabolite of D-methamphetamine), and therefore, if the specimen contains 500 ng/ml of D-methamphetamine and 500 D-amphetamine, one would expect the response to be equal to or greater than that of 1000 ng/ml of the calibrator. It should be noted that for the EMIT d.a.u., which uses D-methamphetamine as calibrator, the response would be greater because of the cross-reactivity of D-amphetamine. In fact, one of the claims made about this kit is that 300 ng/ml of D-amphetamine is equivalent to 1000 ng/ml of D-methamphetamine.

A different situation exists with FPIA. Here the calibrator is D-amphetamine and the cross-reactivity toward D-methamphetamine varies, depending on its concentration. In this case there is an increased response toward mixtures of amphetamine and methamphetamine such that combined concentrations of less than 1000 ng/ml may test positive.

A similar situation exists with the KIMS assay, which was specifically designed to satisfy the federal reporting requirement, which states that in order to report a confirmed positive result for methamphetamine in urine (at or above 500 ng/ml), amphetamine must also be present at or above 200 ng/ml.[7] The assay has little cross-reactivity with pure methamphetamine but enhanced cross-reactivity when amphetamine is present with methamphetamine. This creates two potential scenarios: (1) the assay may detect specimens that contain less than 1000 ng/ml methamphetamine when amphetamine is present and (2) proficiency specimens containing only methamphetamine that are designed to test the laboratories' ability to follow reporting requirements may not produce a positive test result. For practical purposes using donor specimens, the assay has been found to be equivalent to the EMIT assays in detecting positive specimens from methamphetamine users.[21]

An examination of the cross-reactivities of the different tests toward *l*-methamphetamine might lead one to the conclusion that therapeutic use of Vicks Inhaler would not result in a positive response using these kits. Fitzgerald et al.[22] reported concentrations of methamphetamine as high as 6000 ng/ml following use of this decongestant, and these would certainly not be expected to elicit a positive response. He also reported *l*-amphetamine concentrations as high as 455 ng/ml. However, the situation is not as clear as it first seems. The College of American Pathologists, in its Forensic Urine Drug Testing Program, challenged laboratories with a specimen containing approximately 9000 ng/ml of *l*-methamphetamine and 2000 ng/ml of *l*-amphetamine. Greater than 80% of the respondents reported the drugs as positive. Given that the majority, if not all, of the respondents used one of the immunoassays mentioned above, there is likely an additive response in the presence of *l*-methamphetamine and *l*-amphetamine (as there would be for the D-isomers). One obvious outcome of this is that the physician reviewing drug test results should routinely request the separation of stereoisomers by GC/MS.

Methamphetamine can be synthesized by different routes and some of these may lead to the formation of D,*l*-methamphetamine rather than D-methamphetamine.[23] If the donor has used this racemic mixture, then the urine will be less reactive than if he or she had used the pure D-isomer. Under these conditions, 1000 ng/ml of D,L-amphetamine may not trigger a positive response using the immunoassays; the exact response will depend on the amount of methamphetamine and amphetamine present and the additive response of the kit to this mixture.

From this discussion, it is apparent that the exact response of a specimen containing various amounts of amphetamines and varying ratios of the stereoisomers cannot be predicted with any certainty and that the various commercially available immunoassays will react differently. Is this important? It is unlikely that any of these kits are failing to detect the majority of methamphetamine abusers; however, the possibility exists that specimens from some abusers are not being detected and that specimens from users of Vicks Inhaler are being reported as positive (particularly outside of the regulated programs). Without a standardization of the various kits used (for example, requiring them to detect 500 ng/ml of D-methamphetamine and 500 ng/ml of D-amphetamine on a 1:1 basis), the situation will remain as varied as it is today.

Recently there has been significant attention given to the illicit stimulants 3,4-methylene-dioxymethamphetamine (MDMA, also known as "Ecstasy") and 3,4-methylenedioxyamphetamine (MDA, also known as the "love pill"). Currently, HHS regulations do not allow for the testing of these drugs. However, some immunoassays designed for testing for amphetamines possess considerable cross-reactivity to these drugs. Immunoassay kits designed specifically to detect these agents are available as well. For example, some ELISA assays (e.g., Immunalysis) have kits specifically designed for the analysis of either D-methamphetamine (and MDMA) or D-amphetamine (and MDA). Both kits have very low cross-reactivity to the L-isomers of methamphetamine and amphetamine, respectively. However, as the methamphetamine kit has low cross-reactivity to amphetamine and the amphetamine kit has low cross-reactivity to methamphetamine, these types of kits would not be particularly well suited for drug testing under the HHS Guidelines. A similar situation is encountered for several radioimmunoassays as well.

6.1.6.2 Cannabinoids

Marijuana is a popular, potent hallucinogen used primarily for its euphoric effects. The psychoactive constituent in marijuana is δ-9-tetrahydrocannabinol (THC). THC is metabolized to 11-hydroxy-δ-9-tetrahydrocannabinol, which is rapidly metabolized to 11-nor-δ-9-tetrahydrocannabinol-9-carboxylic acid (THCA). THCA may be detected in urine specimens for variable periods of time and potentially for up to several weeks depending on the frequency of use.

Immunoassay techniques are designed to detect cannabinoids in urine specimens utilizing THCA as the target analyte. Other non-cannabinoid analytes usually do not cross-react with the cannabinoid assays. Marinol®, synthetic THC, used for the treatment of anorexia associated with AIDS, will produce positive test results. The original formulation of the EMIT cannabinoid assay was subject to false-positive test results in the presence of ibuprofen. The assay was reformulated in 1986, and is no longer affected by ibuprofen.

Cannabinoid immunoassay cutoff concentrations will affect the percentage of nonconfirmable positive screening results, especially in specimens with urinary concentrations of THCA between 50 and 100 ng/ml. Further, it is difficult to correlate immunoassay response with GC/MS confirmation results for THCA since immunoassay results reflect total cannabinoid metabolite concentration, rather than THCA concentration only.[24]

A problem associated with the interpretation of urine cannabinoid test results is that positive tests may occur from passive inhalation of marijuana smoke. The presence of cannabinoid metabolites in urine specimens due to passive inhalation is a function of environmental conditions, duration and frequency of exposure, and THC content of the smoked marijuana.[25-28] Although passive inhalation studies have shown that individuals exposed to marijuana cigarette smoke under typical conditions would not test positive for cannabinoids, in order to avoid interpretive problems, the federally mandated THCA cutoff concentration was initially set at 100 ng/ml. The cutoff concentration was lowered to 50 ng/ml in 1994.[7] However, passive inhalation studies were performed with marijuana cigarettes containing low concentration of THC (less than 4.0%). Cigarettes seized by law enforcement personnel in recent years have demonstrated THC concentrations exceeding 20%, clearly demonstrating the need to reinvestigate passive inhalation issues.

6.1.6.3 Cocaine (Metabolite)

Cocaine is a potent central nervous system stimulant and local anesthetic. It is rapidly hydrolyzed to benzoylecgonine and ecgonine methyl ester by metabolic and chemical reactions. Benzoylecgonine, the primary cocaine metabolite in urine, is readily detected for 1 to 3 days. Following chronic use of cocaine, benzoylecgonine may be detected for longer periods of time.

Immunoassay techniques are designed to detect cocaine metabolites in urine specimens utilizing benzoylecgonine as the target analyte. Cross-reactivity with other non-cocaine compounds such as lidocaine, benzocaine, and procaine does not occur and will not produce false-positive test results.

In the mid-1980s it was found that Health Inca Tea, which was sold in U.S. health food stores, contained trace amounts of cocaine. The U.S. Food and Drug Administration has since banned the importation of any tea containing residual cocaine. Yet several studies performed with Health Inca Tea and other teas imported from South American countries clearly show that even very low amounts of cocaine in a beverage can result in a positive urine drug test. Health Inca Tea has been reported to contain an average of 4.8 mg of cocaine/bag[29] and between 1.87 and 2.15 mg of cocaine/cup of prepared tea.[30,31] In four subjects ingesting one cup of Health Inca Tea containing 1.87 mg of cocaine, peak urinary benzoylecgonine concentration ranged from 1400 to 2800 ng/ml 4 to 11 h after ingestion, well above the 300-ng/ml regulated immunoassay cutoff concentration.[31] In another study where one subject drank a cup of tea containing 2.15 mg of cocaine, the peak urinary benzoylecgonine concentration of 1280 ng/ml occurred 2 h after tea ingestion.[30] In a study where 6 oz. of Coca-Cola fortified with 25 mg cocaine HCl was consumed by a 165-lb male, peak urinary concentrations of benzoylecgonine (7940 ng/ml) were obtained at 12 h.[32] Urine benzoylecgonine concentrations remained in excess of 300 ng/ml for 48 h.

These studies demonstrate that very small amounts of cocaine, <5.0 mg or about one fifth of a typical intranasal dose (or line), when added to or contained in a beverage, are sufficient to produce a positive drug test in an unsuspecting subject for cocaine metabolites in urine. In such cases, the detection of unique cocaine metabolites, such as methylecgonidine or ecgonidine, produced after smoking crack cocaine, may prove to be useful in refuting alleged oral ingestion of cocaine after a positive drug test for benzoylecgonine; however, testing for these analytes is not currently permitted under federal drug testing guidelines. Instead, the guidelines provided to Medical Review Officers (MROs) state that unknowing ingestion of cocaine is not an alternative medical explanation for the presence of benzoylecgonine in the donor's urine.

6.1.6.4 Opiates

Opiates are potent drugs that are routinely utilized as analgesics to treat moderate to severe pain. Legitimate pharmaceutical sources of opiates include morphine and codeine. Heroin, an illicit substance, is most often abused for production of euphoria and avoidance of withdrawal symptoms. Heroin is rapidly metabolized to 6-acetylmorphine, which is further metabolized to morphine. Codeine is also metabolized to morphine. Opiates may be detected in urine specimens for up to 24 h or for several days, depending on the amount and agent administered.

Immunoassay techniques are designed to detect opiates and their metabolites (primarily morphine, codeine, and their glucuronide metabolites) in urine specimens using morphine as the target analyte. Most opiate assays are generally nonspecific and cross-react with many opiate compounds. Under the federal drug testing program, only morphine, codeine, and 6-acetylmorphine may be reported. In nonregulated testing, other opiates, especially hydrocodone, hydromorphone, and oxycodone, are frequently included as part of the opiate panel, although it should be noted that oxycodone frequently has limited cross-reactivity in general opiate immunoassays. Oxycodone-specific immunoassays are available.

Morphine and codeine have been identified in urine specimens of individuals who have ingested poppy seeds and/or poppy seed cakes. Opiate immunoassays are sufficiently sensitive to detect morphine and morphine conjugates in these specimens days following ingestion. These results are true positives, even though not indicative of opiate abuse. Therefore, in the case of a positive opiate test result, additional testing may be required to differentiate the heroin user from the poppy seed consumer.[25,33] In an attempt to address this issue, in 1997 the mandatory guidelines were changed to increase the initial and confirmatory testing cutoffs for morphine and codeine from 300 to 2000 ng/ml.[34] In addition, the requirement for testing for 6-acetylmorphine (cutoff concentration 10 ng/ml) when morphine is present at concentrations at or above 2000 ng/ml was added. These changes were implemented on December 1, 1998.

6.1.6.5 Phencyclidine

Phencyclidine (commonly known as PCP), a hallucinogen and dissociative anesthetic, is readily detected in urine specimens for up to several days following administration. Immunoassay techniques are designed to detect phencyclidine in urine specimens utilizing phencyclidine as the target analyte. Cross-reactivity with other non-phencyclidine-related compounds is not usually observed except for high concentrations of thioridazine in some assays.[35]

6.1.7 Testing for Other Drugs by Immunoassay

6.1.7.1 Barbiturates

Barbiturates are central nervous system depressants and are utilized as sedative-hypnotic and anticonvulsant agents. There are many barbiturate derivatives that constitute this class of drugs and are classified as ultrashort-, short-, intermediate-, and long-acting agents. Detection of a specific barbiturate in urine depends upon the type of barbiturate ingested, and ranges from 24 h (short-acting) to several weeks (long-acting).

All immunoassay techniques are designed to detect the majority of short-, intermediate-, and long-acting barbiturates in urine specimens using secobarbital as the target analyte. A positive test result indicates the presence of barbiturates in the urine specimen. Additional testing utilizing a chromatographic technique is necessary to identify the specific barbiturate(s) present.

6.1.7.2 Benzodiazepines

Benzodiazepines are central nervous system depressants utilized for their anxiolytic, sedative–hypnotic, anticonvulsant, and muscle relaxant properties. There are many benzodiazepines that constitute this class of drugs, and they are classified as ultrashort-, short-, and long-acting agents. Detection of a specific benzodiazepine in urine depends on the type of benzodiazepine ingested and ranges from 5 to 7 days.

Immunoassay techniques are designed to detect benzodiazepine compounds and their metabolites in urine specimens utilizing oxazepam or nordiazepam as the target analyte. A positive test result requires further testing utilizing a chromatographic technique to identify the specific benzodiazepine compound and/or metabolite(s) present in the specimen. Depending on the immunoassay technique, other benzodiazepines, such as alprazolam, clonazepam, lorazepam, and triazolam, may be difficult to detect.

6.1.7.3 Methadone

Methadone is a synthetic opioid that has analgesic actions and potency similar to morphine. Although its primary use has been in the detoxification and maintenance of heroin addiction, its use is also indicated for treatment of severe pain. This latter use has become much more common in the past few years and reports of methadone abuse and methadone-related fatalities are increasing.[36] Methadone is readily detected in urine specimens for up to 72 h after administration.

Immunoassay techniques are designed to detect methadone in urine specimens using methadone as the target analyte. Although the methadone assays usually do not cross-react with compounds structurally unrelated to methadone, the presence of metabolites in urine specimens from individuals administered L-alpha-acetylmethadol (LAAM), a long-acting methadone analogue, may produce positive test results.

6.1.7.4 Methaqualone

Methaqualone is a central nervous system depressant. Although methaqualone was a popular drug during the 1970s and early 1980s, its use has been replaced by other drugs such as benzodiazepines. Methaqualone is not currently available legally in the U.S. Following administration, methaqualone can be detected in urine specimens for up to 14 days. Immunoassay techniques are designed to detect methaqualone in urine specimens using methaqualone as the target analyte. Cross-reactivity with other non-methaqualone-related compounds is not usually observed.

6.1.7.5 Propoxyphene

Propoxyphene is a synthetic opioid that has analgesic actions and potency similar to codeine. Its use is indicated for the treatment of moderate to severe pain. Propoxyphene and its primary metabolite, norpropoxyphene, are present in urine specimens for several days. Immunoassay techniques are designed to detect propoxyphene and norpropoxyphene in urine specimens utilizing propoxyphene as the target analyte. Cross-reactivity with other non-propoxyphene-related compounds is not usually observed.

6.2 CONFIRMATORY TESTING

Before the introduction of the HHS Guidelines[6] there was considerable discussion among forensic toxicologists regarding the appropriate analytical procedures to be used to confirm immunoassay results. There was a recognition that these had to be based on a different chemical principle than immunoassay, but there was not a consensus that this needed to be GC/MS. However, the Guidelines mandated that GC/MS be used to confirm presumptive positives from immunoassay testing and its use was further incorporated into the College of American Pathologist's FUDT inspection program. It has now become the standard of practice in workplace drug testing to confirm all initial immunoassay results by GC/MS. Common cutoff concentrations are listed in Table 6.3.

Not only was there considerable discussion on the need for GC/MS, there was also considerable debate about the suitability of selected ion monitoring for the identification of drugs and/or their metabolites. Several well-respected forensic toxicologists argued that full scan GC/MS was necessary before a specimen could be reported as positive. Although this argument is still heard today, there is a broad consensus of opinion that selected ion monitoring is an accurate and reliable procedure for identification of these substances. The analytical principles of GC/MS have been well covered by other

Table 6.3   GC/MS Cutoff Concentrations

Drug Group	Common Analyte(s)	Cutoff Concentration (ng/ml)
Amphetamines	Amphetamine, methamphetamine	300 or 500[a],*
Barbiturates	Amobarbital, butalbital, pentobarbital, phenobarbital, secobarbital	200 or 300
Benzodiazepines	alpha-Hydroxyalprazolam, nordiazepam, oxazepam, temazepam	100, 200, or 300
Cannabinoids	THCA	10 or 15*
Cocaine	Benzoylecgonine	150*
Methadone	Methadone	100 or 300
Methaqualone	Methaqualone	300
Opiates	Codeine, morphine (6-acetylmorphine)	300 or 2000* (10*)
Phencyclidine	Phencyclidine	25* or 75
Propoxyphene	Propoxyphene and/or norpropxyphene	300

[a] 200 ng/ml amphetamine must be present to report methamphetamine positive.
* Cutoffs used in the SAMHSA Guidelines and the regulated programs.

authors.[37,38] The HHS Guidelines[6,7] required certified drug testing laboratories to be inspected every 6 months, and it is this inspection process that has driven the improvement in the confirmation procedures. This section examines some of the quality assurance/quality control procedures that have become common practice in certified drug testing laboratories and briefly discusses some analytical issues.

6.2.1 Quality Assurance

To provide adequate control and verification of the analytical process, laboratories must implement appropriate policies and procedures regarding GC/MS analysis. This section is intended to discuss the quality control and quality assurance requirements pertinent to the regulated drug-testing laboratory, although many of the components specified below are directly applicable to any laboratory performing drug testing in biological specimens. Much of the information in this section has been condensed from a comprehensive review of commonly practiced quality control and quality assurance procedures in forensic urine drug testing laboratories.[39]

6.2.1.1 Method Validation

Method validation is a process of documenting or proving that an analytical method is acceptable for its intended purpose. For analytical methods to be implemented in laboratory-based regulatory drug testing programs, the laboratory must be able to demonstrate that the chosen analytical method has the ability to provide *accurate* and *reliable* data. At a minimum, the key assay characteristics to be established and evaluated should include the accuracy, precision, linearity, specificity, sensitivity, carryover potential, and ruggedness of the analytical method. Additional characteristics to be evaluated may include the stability of the analyte under various analytical and storage conditions, identification and concentration of the internal standard(s) for the method, validation of use of partial (diluted) sample volumes, and estimated recovery of the analyte from the matrix.

Accuracy and Precision

Two of the most important assay characteristics to be determined during method validation are accuracy and precision. Together, accuracy and precision determine the error of an analytical measurement. The accuracy of a method refers to the closeness of the measured value to the true value for the sample. Precision, on the other hand, refers to the variability of measurements within a set. It is most often used to demonstrate scatter or dispersion between numeric values in a set of measurements that have been determined under the same analytic parameters.

The accuracy and precision of an assay can be determined by comparing test results utilizing laboratory-prepared standards and controls with those obtained with an established reference method or by analysis of standard reference materials. Secondary checks may involve reanalysis of specimens whose concentrations are already known to be accurate (e.g., patient or proficiency samples analyzed by reference laboratories using a validated method).

Accuracy is generally expressed as the percentage difference from the actual value. The assessment of accuracy must be carried out on mean values that have been calculated from replicate measurements of reference materials containing known concentration of analyte. It has been recommended that accuracy be assessed using a minimum of nine determinations over a minimum of three concentrations (i.e., 3 concentrations/3 replicates each).[40,41]

Accuracy acceptability ranges in forensic urine drug testing laboratories should not exceed 20% (by convention) of the target concentration. Many laboratories routinely use lower ranges, such as 10%. It should be noted that the acceptable accuracy range selected for initial method validation may differ from that selected for routine use (batch acceptance criteria). For example, a laboratory may require that accuracy be within 10% of the known concentration during method validation, and then choose to increase the acceptable range to 20% for routine daily analysis.[42]

Precision of an analytical method is usually assessed in two ways: analysis of multiple measurements during a single analytical run (within-run precision) and analysis of single, or mean, measurements over many runs (between-run precision). Precision is expressed as the percentage relative standard deviation or coefficient of variation (CV). Within-run precision can be considered a measure of the precision of an analytical method under optimal conditions. The between-run precision, however, is a better representation of the precision one might observe during routine use of the assay. The precision of an assay should be determined at concentrations below, at, and above the assay cutoff concentration for that assay. This can be accomplished by repeated analyses of quality control samples on a within- and between-batch basis. Precision should be determined not only when validating the methods, but on an ongoing basis. The SAMHSA guidelines currently require annual review of accuracy and precision. Generally, within-run and between-run CV values of less than <15% are considered acceptable.[43–45] However, greater variability is to be expected as analyte concentrations approach the limit of detection (LOD) of the analytical method and the laboratory might choose to increase the acceptability criterion to 20% at its lowest measured concentrations.[46]

Linearity and Sensitivity of the Analytical Method

The linear range of a method should be established during initial assay characterization and periodically thereafter with specimens containing drug analytes over a wide range of concentrations. In forensic urine drug testing, linearity should be established via evaluation of a plot of instrument response (y) as a function of analyte concentration (x) using least-squares analysis.[47,48] Using a series of calibrators that have been prepared at various known concentrations of analyte, the individual concentrations of each calibrator are "reverse calculated" using the regression line generated. Acceptable linearity is demonstrated when the correlation coefficient exceeds a defined value, such as 0.990, and quantitative concentration of each point falls within ±20% of the target value. For purposes of forensic urine drug testing, the limit of quantitation and upper limit of linearity of the assay are typically defined as the lowest and highest concentrations, respectively, where all quantitative *and* qualitative quality control criteria are met. The limit of detection of the assay is the lowest concentration for which qualitative quality control criteria are met. Other approaches (e.g., signal/noise ratios) may be used to establish these parameters and have been reviewed elsewhere.[39]

The laboratory uses this information to establish the range, or concentration interval, which it will routinely use for analysis of samples. Validating the method over a wider range than that used in daily practice provides increased confidence that the routine standard concentrations are well removed from nonlinear response concentrations.

Specificity of the Analytical Method (Interference Studies)

Specificity refers to the ability of the analytical method to accurately measure an analyte response in the presence of all potential sample components. All methods should at a minimum be investigated for potential interference by endogenous matrix components, as well as common compounds that are structurally similar to the analyte of interest.

The potential interference of endogenous urine components with the assay is most frequently assessed by evaluation of urine specimens from several sources (donors) that are known to be drug-free for the analyte of interest. Assessment of interference from structurally related compounds can be determined by fortification of urine with high concentrations (e.g., 1 mg/ml) of potentially interfering analytes and cutoff concentrations of target analytes, or with concentrations of analyte that are expected under therapeutic conditions. For example, possible interference with the measurement of amphetamine and methamphetamine may occur due to the presence of sympathomimetic amines such as ephedrine, pseudoephedrine, phenylpropanolamine, and phentermine.[49] Further, interference with the measurement of morphine and codeine due to the presence of opiate metabolites and synthetic 6-keto-opioids such as dihydrocodeine, hydromorphone, hydrocodone, oxymorphone, and oxycodone has also been described.[50]

The problem of interfering substances may be addressed by employing more selective extraction methods, chromatographic separations, or detection methods. For example, to eliminate potential false-positive amphetamine/methamphetamine results due to the presence of other sympathomimetic amines, aliquots of specimens can be treated with a solution of 0.035 M sodium periodate at room temperature, then subjected to extraction. In the presence of periodate, α-hydroxyamines undergo oxidative cleavage removing the potential interferent.[51] It is recommended that periodate oxidation be conducted at pH 7 or lower to prevent possible formation of low levels of amphetamine from extremely high levels of methamphetamine that may be present in the specimen.[52]

Carryover

The term *carryover* is used to refer to the contamination of a sample by a sample analyzed immediately prior to it.[53,54] In the urine drug testing laboratory, the term *carryover limit* is used to delineate the concentration of analyte in a sample above which contamination may reasonably be expected to occur. There is at least one common approach to performing such studies that involves the analysis of standards prepared at increasingly higher concentrations of analyte, preferably reflecting the highest concentrations that a laboratory typically encounters during routine analysis of samples. Each standard should be injected separately, followed by injection of a blank or solvent to determine if a signal (response) characteristic of the analyte is present in the sample above a pre-established limit (typically the limit of detection). Once the concentration at which carryover occurs is determined, the laboratory establishes its carryover limit at the next lowest concentration that did *not* have evidence of carryover in the blank or solvent. More precisely, upon completion of carryover studies, a laboratory should define the range of analyte concentrations at which carryover does not occur. The laboratory should also ensure that the quantitative value for the carryover limit established in the carryover study falls within the linearity of the assay to ensure that the quantitative value is accurate.

It is also advisable to evaluate carryover of an assay on each instrument system, including autosamplers, on which the method is to be performed, although there is no general consensus on this issue. This is to ensure that the established carryover limit is properly applied to data obtained on each system routinely used in the laboratory. It is important that criteria be established for evaluating the acceptability of solvent blanks or negative quality control samples that have been inserted to assess possible carryover.[55] If carryover is suspected, a potentially contaminated specimen should be reextracted, rather than reinjected, because the extract vial may have been contaminated.

Other Factors

Other factors include selection of a derivatizing reagent, selection of the internal standard(s) for the assay, and selection of ions to monitor for selected-ion monitoring or full-scan analysis. These topics are considered in Section 6.2.4 (Analytical Issues).

Another important parameter to assess during the method validation phase is that of stability of the analyte. This includes stability of stock solutions of analyte, as well as stability of the analyte in biological matrix. Stability studies will typically be performed to assess stability under different temperatures (storage conditions) and different lengths of time (in-process stability and long-term stability). It is recommended that the laboratory at least establish analyte stability under its own anticipated storage and processing conditions. While there are many different approaches to the performance of stability studies, a common approach is to use quality control materials prepared at known concentrations to assess stability.[43]

6.2.2 Calibrators and Controls

6.2.2.1 Assay Calibration

In forensic urine drug testing laboratories, assay calibration is necessary to determine whether an analyte is present in an unknown sample at or above a preestablished administrative cutoff value, as well as to determine the accurate quantitative concentration of the analyte under certain circumstances.[56] Calibrators are prepared from a mixture of different amounts of known concentrations of analyte standards with a fixed amount of internal standard. The two most common approaches to assay calibration in forensic urine drug testing laboratories are multipoint calibration curves and single-point calibrations.

Multipoint calibration curves must include calibrators that bracket the cutoff concentration. Although many laboratories include a calibrator at the cutoff concentration, it is not required.[56] For each calibrator, specific ion-abundances are measured for the analyte and internal standard. A ratio is calculated for each calibrator and a calibration curve is generated using a simple least-squares regression model. Generally accepted criteria for acceptable linearity using deuterated internal standards for calibration include a correlation coefficient of at least 0.990, and a y-intercept close to zero, although slightly positive or negative y-intercept values are also acceptable. For assay calibration, the laboratory must also establish whether the regression line will be forced through the origin (a "no-intercept" model) or will not be forced through the origin (an "intercept" model). Selection of a no-intercept model implies that the assay limit of detection is *zero*, which is usually not the case for GC/MS urine drug testing assays. Therefore, in most instances regression through the origin should not be used in assay calibration. Finally, the calculated concentration of all standards against the constructed calibration curve should be within ±20% of their respective target concentrations.

Single-point calibrations utilize a calibrator containing analyte at the assay cutoff concentration. Laboratories then include quality control materials at concentrations below, at, and above and below the cutoff concentration to demonstrate linearity. The quantitative results for the quality control materials used in single-point calibration must fall within 20% of their respective target concentrations.

During assay calibration, the laboratory also establishes acceptance ranges for retention time and ion ratios. Acceptance ranges are established to evaluate each calibrator, control, and unknown specimen in the batch. For single-point calibration assays, the acceptable limits are determined from the calibrator at the cutoff. For multipoint calibration curves, acceptable limits may be determined from either the calibrator at the cutoff *or* from the average of all the calibrators analyzed. The laboratory must apply the acceptable ranges consistently to all calibrators, controls, and specimens in the batch. (See Section 6.2.3, Criteria for Designating a Positive Test Result.)

6.2.2.2 Chromatographic Performance

The chromatographic performance of an assay should also be assessed before the analysis of specimens. This is achieved readily by the injection of an unextracted performance standard including analyte(s) and internal standard(s). Use of an unextracted standard removes sources of variation due to the extraction procedure and matrix interferences. In addition to permitting evaluation of the quality of the chromatographic system, the analysis of an unextracted standard serves to verify that the analyte(s) of interest elute at the expected retention time, that all MS acquisition windows are appropriately set, and that no unexpected adsorptive system losses have occurred. Evaluation criteria in the standard operating procedure manual should include a thorough description of peak shape, resolution, and signal abundance requirements. A complete review of methods for evaluation of chromatographic performance is beyond the scope of this chapter; however, acceptable alternative criteria for evaluation of chromatographic acceptability can be found in many textbooks and other reference materials.[54,57–61]

6.2.2.3 Quality Control

Under the HHS guidelines each assay batch must include a minimum total of 10% open and blind positive and negative quality control samples in an appropriate urine matrix.[6,7] Quality control samples may be purchased commercially or prepared from a different source or lot of standard material than that used to prepare calibrators. The target concentration at one control must be within approximately 125% of the cutoff concentration; other controls should be prepared at appropriate concentrations in order to regularly access accuracy below and/or above the assay cutoff concentration. It is recommended that a highly concentrated control sample be diluted in a similar manner as diluted specimens during aliquoting in order to verify the accuracy of the dilution technique, if one is routinely used to prepare presumptive positive specimens for confirmation.[56]

To evaluate the efficiency of the hydrolysis process, cannabinoid and opiate control samples containing conjugated THCA and morphine-glucuronide, respectively, should be assayed. Negative urine samples spiked with these reference materials can be prepared, or as an alternative, hydrolysis control samples can be prepared from a combined urine pool of previously confirmed specimens (which would otherwise be discarded). An additional quality control that includes potentially interfering substances (such as a control-containing phenylpropanolamine, ephedrine, phentermine, and/or pseudoephedrine) is also recommended.

Verification of Quality Control Materials

Prior to the use of reference materials to prepare calibrator, control, or internal standard material in the laboratory, the laboratory is required to verify that its chemical identity is correct and that it is of acceptable purity and concentration *independently* from the supplier. At a minimum, most laboratories perform a full-scan GC/MS analysis to verify the chemical identity and purity of the material and compare the obtained spectra with that of available library spectra to determine that significant impurities are not present in the material which might interfere with the method. The isotopic purity of the internal standard can also be verified with this procedure. Additional methods for verification of chemical identity and purity may involve measurement of physical constants, such as melting point or refractive index, as well as use of other analytical techniques (HPLC, IR, NMR, TLC, or UV/VIS) to detect nonpolar or nonvolatile impurities.[54] Verification of concentration is most often evaluated indirectly by preparation of calibrators or controls at known concentrations and analysis in routine batches. The laboratory must establish specific evaluation criteria for reference materials, such as spectral match requirements, percent isotopic purity required, and quantitative results. The laboratory must retain documentation of all verification procedures performed.[56] The laboratory may then use the reference material to prepare calibrators for controls for routine use. Of course, these new calibrators and controls must then be themselves validated for concentration (e.g., ±20% of the target concentration) prior to routine use.

Evaluation of Quality Control Results

In forensic urine drug testing, the most common approach to the evaluation of quality control results is through the use of a fixed-criterion quantitative acceptance range, usually ±20% of the target concentration. The laboratory's standard operating procedure (SOP) manual must thoroughly describe the quality control evaluation criteria to be used and must include a policy for the required course of corrective action if quality control sample results fail to meet acceptance criteria. To assess laboratory performance, all control data, including out-of-limit data, should be recorded in the quality control log in Levey–Jennings or Shewart chart format.[56,62–67] Out-of-limit data should include documentation of required corrective action. It may be acceptable to reinject a quality control sample one additional time. If the results are still unacceptable, other minimally acceptable protocols include (1) reinjection of calibrators, followed by reprocessing of *all* quality control samples and routine specimen data against the new calibration, *if* the time since the last injection is not excessive and the instrument has not been retuned; (2) reinjection of all calibrators, quality control samples, and routine specimens; and (3) acceptance of negative test results that are less than the LOD and reextraction of all other specimens in the batch.

6.2.3 Criteria for Designating a Positive Test Result

6.2.3.1 Qualitative Criteria

Criteria for designating a positive test result include chromatographic and spectral identification. Chromatographic identification of an analyte requires comparison of the retention time of the specimen with that of a calibrator at the cutoff or the average of multiple calibrators. Generally, the retention time of the analyte should be within ±2% of the retention time as established by the calibrator(s).[56,68,69]

Further, identification of an analyte requires comparison of ion ratios or full-scan mass spectra, respectively. Acceptable ion ratios for the analyte and its corresponding internal standard are usually calculated using ion abundance data obtained for the cutoff calibrator or by determining the mean ion ratios for all calibrators. It has been demonstrated statistically that while full-scan mass spectrometric provides the maximum confidence for analyte identification, a minimum of three structurally significant ions generated under electron ionization conditions appear to provide adequate information for an identification.[68,70] If the regulated urine drug testing laboratory uses electron ionization, it is currently required to use a minimum of two ion ratios for identification of the analyte and at least one ion ratio for the internal standard.[56]

The ion ratios for the analyte and its corresponding internal standard obtained for the unknown should not differ by more than ±20% of the target ion ratio and acceptance criteria must be uniformly applied to all specimens within the batch.[62] The establishment of this 20% acceptance criteria for ion abundance ratios has been determined to be appropriate based on ion statistics.[68–70] Different ion ratio criteria cannot be applied to different specimens, calibrators, or controls within the batch.

6.2.3.2 Quantitative Criteria

Regardless of the detection technique, to be designated as positive, the measured concentration of a specimen must be equal to or exceed the established assay cutoff concentration. Quantitative results around the cutoff must be truncated, rather than rounded up to the nearest whole number, so that the statistical bias is toward a "negative" result. Specimens directed for GC/MS retest analyses are not subject to cutoff concentrations and are reported as reconfirmed if the concentration is equal to or greater than the LOD of the method. The specimen retest must also meet all other criteria for designating a positive test result.

If the specimen concentration exceeds the upper limit of linearity for the assay, quantitative reports must state that the concentration of the analyte is "greater than *the established upper limit of linearity*" or be reanalyzed using an appropriate dilution. Also, all criteria for designating a positive test result must be satisfied, including chromatographic performance, ion ratios, and retention time data.

6.2.3.3 Data Review

All data must be thoroughly reviewed by a minimum of two individuals to verify compliance with the methods specified in the laboratory's procedure manual and to identify clerical and analytical errors. Batch acceptance criteria include within-range standards and controls and acceptable MS tune, chromatographic performance, ion abundance (adequate signal-to-noise ratio), and ion ratios or mass spectral match quality. Also, the laboratory may choose to monitor the consistency (reproducibility) of the ion abundance for the internal standard to ensure that it has been added appropriately, and at the correct concentration, to each calibrator, control, and unknown in confirmatory analyses. Furthermore, an acceptable calibration curve must be obtained, a lack of carryover must be demonstrated, and chain-of-custody documentation must be in order.

The SOP must also address the handling and reporting of results when duplicate extracts are assayed, or diluted and undiluted extracts are analyzed. Acceptance criteria for duplicates must specify minimum correlation of quantitative results, usually ±20% if both results fall within the assay's linear range. Both of the results must be equal to or greater than the mandated cutoff concentration. If reporting quantitative results, the least diluted sample within the assay's linear range is typically reported. Review of specimen data should include comparison of the initial immunoassay response with the GC/MS result, an evaluation of chromatographic performance including presence of interfering or extraneous peaks, retention time, minimum ion abundance, ion ratios or mass spectral match quality, quantitation, extraction efficiency, potential for carryover, and chain-of-custody documentation.

6.2.4 Analytical Issues

It is far beyond the scope of this section to discuss specific analytical procedures for confirmation of drugs. Methods are readily available in the literature and have been described for urine as well as other matrices. However, it is important to discuss a few of the specific analytical issues that the laboratory will need to face when performing confirmation testing for drugs of abuse by GC/MS.

6.2.4.1 Selection of a Derivative

Selection of a suitable derivative is a critical component of an assay when required. There are at least three major reasons for using a derivatized compound. First, the analyte can often be made sufficiently volatile to allow its introduction to the mass spectrometer by gas chromatography, permitting optimal separation of the analyte from possible interfering substances. Second, the stability of the analyte during storage, isolation, and thermal volatilization can be enhanced via formation of the derivatized product. Third, the increase in molecular mass resulting from derivatization on the fragmentation may be beneficial, providing ions which, by virtue of their higher mass, are more specific for the analyte.[71,72] The ideal derivatization procedure should be convenient and rapid to perform, form a consistent and stable product in high yield, require small volumes, be selective for the analyte of interest, be safe to handle, and should not form by-products that interfere with the analysis.[73–77]

Table 6.4 summarizes the numerous choices of derivatizing reagents available to today's drug testing laboratory. Table 6.4 does not include all the analytes that can be included under each drug group. For example, the amphetamines assay can be modified to include MDMA, MDA, and MDE, the benzodiazepines to include α-hydroxyalprazolam, α-hydroxytriazolam, diazepam, and N-desalkylflurazepam, and the opiates to include hydrocodone and hydromorphone. Detailed information about these procedures is available in the references listed. However, there are issues associated with a number of the assays that do deserve discussion.

Amphetamines

In the early 1990s, it was discovered that during certain GC/MS procedures and under certain conditions, methamphetamine can be formed from large concentrations of ephedrine (or pseudoephedrine). Consequently, HHS introduced a requirement that for a specimen to be reported as positive for methamphetamine, it must not only contain at least 500 ng/ml of methamphetamine but also at least 200 ng/ml of amphetamine.[7] This prevented the reporting of false-positive meth-

Table 6.4   Derivatization Procedures

Drug Group	Drugs Commonly Included	Derivatizing Procedure	Ref.
Amphetamines	Amphetamine Methamphetamine	Acylation Silylation	23, 84–86
Barbiturates	Amobarbital Butabarbital Pentobarbital Phenobarbital Secobarbital	Alkylation	87
Benzodiazepines	Nordiazepam Oxazepam	Silylation	88–90
Cannabinoids	THCA	Acylation/esterification Alkylation Silylation	91–94
Cocaine	Benzoylecgonine	Alkylation Silylation	95, 96
Opiates	Codeine Morphine 6-Acetylmorphine[a]	Alkylation Acylation Silylation	97–100

[a] See discussion in text.

amphetamines as there was no evidence of the formation of amphetamine during the GC/MS process. This rule does not apply to nonregulated testing, and there may be noncertified laboratories that have not incorporated the reporting rule into their procedures.

A more reliable procedure for eliminating the potential for false-positive results is to incorporate a pre-oxidation step into the extraction.[51] In this protocol, oxidation with sodium periodate destroys hydroxylated sympathomimetics before the extraction, thereby preventing their conversion to methamphetamine during derivatization or GC/MS. If this procedure is used, then good laboratory practices require the laboratory to analyze a control containing the amphetamines together with high concentrations of the sympathomimetics as part of their standard protocols.

In the initial discussion of immunoassay testing for amphetamines, it was mentioned that under certain conditions L-methamphetamine (together with L-amphetamine) will cause a positive amphetamines response. Because of this, it is important that laboratories incorporate procedures to separate the stereoisomers into their protocols. Such procedures are based on derivatization with an optically pure derivative of an amino acid.[22] Interpretation of these GC/MS results is usually straightforward: specimens from Vick's Inhaler users contain greater than 90% of the L-isomer and those from users of D-methamphetamine contain greater than 90% of the D-isomer.

Benzodiazepines

Over the past 5 years numerous publications have focused on this GC/MS assay for benzodiazepines. Historically, it was possible to convert the parent drugs and their metabolites to benzophenones by acid hydrolysis;[78] however, this procedure does not allow for the confirmation of α-hydroxyalprazolam or α-hydroxytriazolam, metabolites of two of the newer, and most commonly prescribed, benzodiazepines. Using this procedure it is also not possible to differentiate the use of certain benzodiazepines, leading to potential interpretative difficulties.

For these reasons, acid hydrolysis has fallen out of favor and been replaced with procedures involving GC/MS analysis of the parent drug or their metabolites. Because a number of benzodiazepines and their metabolites are metabolized by glucuronidation, it is necessary to perform a hydrolysis step before extraction. Unlike the hydrolysis used for opiates, which can be either acid or enzymatic, only enzymatic ones can be used in the extraction of the benzodiazepines. Because of the polar nature of the benzodiazepines it is necessary to derivatize them; normally the trimethylsilyl derivatives are formed.

The major issue surrounding benzodiazepine testing is the cutoff to be used for confirmation. Historically, either a 200- or 300-ng/ml cutoff has been used; however, this is inappropriate for many of today's benzodiazepines. For example, to confirm many of the immunoassay positives resulting from alprazolam or flunitrazepam use it is necessary to lower this cutoff to 100 ng/ml. As the potency of these drugs increases it will be necessary to constantly monitor whether the confirmation (and the screening) assays are appropriate for their detection. Even in today's environment it is surprising how many laboratories continue to confirm only nordiazepam and oxazepam.

Opiates

Although morphine and codeine have been confirmed by GC/MS for a number of years, this assay continues to cause laboratories problems. One of the underlying issues is whether to use acid or enzymatic hydrolysis. Although studies have shown that acid hydrolysis is more efficient in releasing the parent drug from its conjugate, it has the disadvantage of resulting in a "dirty" hydrolysate that can cause problems in either solid-phase or liquid-liquid extractions. For this reason (as well as for safety ones) most laboratories use enzymatic methods, and with the correct choice of enzyme and conditions, hydrolysis can occur quickly and efficiently. Finally, acid hydrolysis will destroy 6-acetylmorphine in a specimen whereas enzymatic hydrolysis is more "gentle." However, most laboratories utilize a different analytical procedure altogether when testing for the presence of 6-acetylmorphine, as hydrolysis is not required and there are other issues relating to its analysis (see below).

A second issue associated with opiate confirmation is the separation and identification of hydrocodone and hydromorphone in the presence of morphine and codeine. This problem is confounded by norcodeine. One of the most common procedures used involves silylation as the derivatization procedure. When this technique is used, the separation of the trimethylsilyl derivatives of codeine from hydrocodone and those of morphine and hydromorphone can be difficult, particularly when it is required to have greater than 90% resolution. It is true that the derivatives can be clearly identified by choosing the appropriate ions to monitor; however, it is still necessary to use common ions. Under normal conditions, the separation of trimethylsilyl derivatives of morphine and norcodeine is almost impossible to achieve.

Therefore, the confirmation of morphine after codeine use can be impractical if the laboratory uses silylation as a derivatization. Under normal conditions, when the donor has only taken codeine, this may not be important; however, if the donor has taken heroin and codeine a negative report for morphine is extremely misleading. The easiest way to resolve this problem is to use a different derivatizing procedure. Either acylation or acetylation is suitable; however, acyl derivatives are notoriously unstable. Acetylation with acetic anhydride or propionic anhydride is recommended. Although it is more time-consuming from an extraction point of view, the derivatives are well separated chromatographically.

For the analysis of 6-acetylmorphine, silylation is the method of choice although other methods have been used. It is obviously inappropriate to use acetylation, as this will acetylate both 6-acetylmorphine and morphine back to heroin; therefore no positive morphine specimens would ever be negative for 6-acetylmorphine! As the regulated laboratory is required to detect 6-acetylmorphine at a cutoff concentration of 10 ng/ml compared with 2000 ng/ml for morphine and codeine, linearity considerations make simultaneous analysis for the determination of morphine, codeine, and 6-acetylmorphine impractical.

6.2.4.2 Selection of an Internal Standard

The selection of a suitable internal standard is highly linked to the appropriate selection of a derivative (if needed) for the assay as well the particular ions to be monitored for analyte identification and quantitation. An ideal internal standard behaves identically to the analyte throughout the extraction, chromatographic separation, and ionization processes. Deuterium-labeled analogues are the most frequently used stable isotope internal standards in urine drug testing laboratories. Several factors are important to consider when selecting the deuterium isotope-labeled internal standard to be used in an assay. The isotope should not undergo exchange under any of the conditions under which it will be used, such as the extraction, derivatization, or chromatographic separation procedures, as well as the mass spectrometer source.[79]

In addition, the isotope must be stable under routine storage conditions, so that exchange of deuterium and hydrogen does not occur.[80] The isotopic variant selected should have a molecular weight three or more mass units greater than the unlabeled compound because the naturally occurring heavy isotope content of organic compounds in general produces ions of significant intensity at one or two mass units above each carbon-containing compound in the analyte's mass spectrum.[79,81–83] It is therefore critical that the isotopic variant is of high purity (>99%) to prevent interference with the analyte of interest during the analysis. Also, the labeling of the isotope should be in such a manner that it is located in the molecular structure so that, after the fragmentation or ionization process, a sufficient number and intensity of high-mass ions that retain the label are present and will not interfere with the intensity measurement of the corresponding ions derived from the analyte.[83] The laboratory must carefully evaluate the concentration of internal standard used in its assay to ensure that there is no contribution to analyte signal itself. Generally, ions of high, even mass-to-charge ratios have fewer possible origins and are therefore more likely to be characteristic of a particular analyte. It is recommended that laboratories using selected-ion monitoring utilize at least three characteristic ions for the analyte of interest, and a minimum of two characteristic ions for the internal standard.[56]

6.2.4.3 Selection of Ions

There has been considerable discussion regarding the choice of ions to monitor and whether they are suitable for identification purposes. For most analytes of interest this is not a problem; however, for some it is. Analytes that clearly fall into the first category include benzoylecgonine, the opiates, and THCA.

Amphetamines

Unless amphetamine and methamphetamine are derivatized, their base peaks in the electron impact mode are 58 and 44 m/z, respectively. These two masses are certainly not diagnostic as they represent a dimethylamine and a methylamine fragment, respectively. Second, monitoring such low ions can lead to a loss of chromatographic selectivity because of the potential for co-extracted interfering compounds with low mass ions. Derivatization with a perfluoroacyl reagent (e.g., heptafluorobutyric anhydride [HFBA]) will result in ions of considerably higher mass and generally much "cleaner" ion chromatograms. However, there is still need for caution in the GC/MS confirmation of amphetamines because of the potential for interferences from other sympathomimetics as was previously discussed.

Barbiturates

The barbiturates (amobarbital, butabarbital, butalbital, pentobarbital, phenobarbital, and secobarbital) are frequently included in nonregulated drug testing programs. Their confirmation presents some difficulty because of the similarity of electron impact mass spectra of certain barbiturates and the absence of a clearly defined third ion for the analyte. For example, the mass spectra of amobarbital, butabarbital, and pentobarbital include two major ions, at 141 and 156 m/z. A similar situation exists with butalbital and secobarbital (167 and 168 m/z). In addition these five barbiturates elute closely together on the majority of capillary columns used today. Phenobarbital has two distinctly different ions (and is well separated chromatographically) at 204 and 232 m/z. Therefore, laboratories performing this confirmation without a derivatization step are, in essence, basing their identification of the barbiturates more on retention time than on the ions monitored. One solution to this problem is to derivatize the barbiturates before GC/MS. This results in more diagnostic ions and improves the chromatographic separation. In our laboratory, alkyl derivatives are formed using Meth Elute.

Benzodiazepines

As previously discussed, GC/MS confirmation of this group of drugs is difficult; however, one of the issues with developing a procedure for these drugs is the choice of ions to monitor. All of the benzodiazepines contain a halogen (fluorine, chlorine, or bromine). If the choice is to monitor a fragment containing chlorine or bromine, the laboratory should be careful not to choose a second fragment that is simply one containing a chlorine or bromine isotope.

Methadone and Propoxyphene

Assays for these two drugs are based on confirming the presence of the parent drug, and are faced with the same problems as those designed to confirm the presence of underivatized amphetamine and methamphetamine: the base peak in the electron impact mass spectra represents an alkyl amine moiety. It is therefore necessary to choose ions of low intensity for monitoring purposes, or to establish assays based on the metabolites (EDDP for methadone and norpropoxyphene for propoxyphene).

Phencyclidine

Most confirmation assays for this drug monitor the following ions: 200, 242, and 243 m/z. Obviously 243 m/z is simply the carbon isotope fragment corresponding to the 242 m/z ion. However, in this case there are no other suitable diagnostic ions. Fortunately, the 243 ion arises from a different molecular fragment and does not arise solely as the result of an isotope of the 243 ion.

ACKNOWLEDGMENTS

Portions of this chapter were adapted from:

Peat, M. and Davis, A.E., Analytical considerations and approaches for drugs, in *Handbook of Drug Abuse*, 1st ed., Karch, S.B., Ed., CRC Press, Boca Raton, FL, 1998, chap. 10.3.
Goldberger, B.A., Huestis, M.A., and Wilkins, D.G., Commonly practiced quality control and quality assurance procedures for gas chromatography/mass spectrometry analysis in forensic urine drug-testing laboratories. *Forensic Sci. Rev.*, 9, 59, 1997.
Goldberger, B.A. and Jenkins, A.J., Drug toxicology, in *Sourcebook on Substance Abuse: Etiology, Epidemiology, Assessment, and Treatment*, Allyn & Bacon, Boston, 1999, 184–196.

REFERENCES

1. 1992 National Household Survey on Drug Abuse. Advance Report 3. Substance Abuse and Mental Health Services Administration (SAMHSA), Office of Applied Studies, DHHS, Rockville, MD, 1993.
2. Monitoring the Future Survey. College Students. National Institute on Drug Abuse, DHHS, Rockville, MD, 1993.

3. Monitoring the Future Survey. High School Students. National Institute on Drug Abuse. DHHS, Rockville, MD, 1993.
4. Anthony, J.C., Eaton, W.W., Garrison, R., and Mandell, W., Psychoactive drug dependence and abuse: more common in some occupations than others. *J. Employee Assistance Res.*, 1, 148, 1992.
5. Harris, M.M. and Heft, L.L., Alcohol and drug abuse in the workplace: issues, controversies and directions for future research. *J. Manage.*, 18, 239, 1992.
6. Mandatory Guidelines for Federal Workplace Testing Programs. *Fed. Register*, 53, 11970, 1988.
7. Mandatory Guidelines for Federal Workplace Testing Programs. *Fed. Register*, 59, 29908, 1994.
8. Gorodetzky, C.W., Detection of drugs of abuse in biological fluids, in *Handbook of Experimental Pharmacology*, Born, G.V.R., Eichler, O., Farah, A., Herken, H., and Welch, A.D., Eds., Springer-Verlag, Berlin, 1977.
9. Wu, A.H.B., Near-patient and point-of-care testing for alcohol and drugs of abuse. Therapeutic Drug Monitoring and Toxicology In-Service Training and Continuing Education Program, American Association for Clinical Chemistry, 16, 227, 1995.
10. Jenkins, A.J., Mills, L.C., Darwin, W.D., Huestis, M.A., Cone, E.J., and Mitchell, J.M., Validity testing of the EZ-SCREEN® cannabinoid test. *J. Anal. Toxicol.*, 17, 292, 1993.
11. Jenkins, A.J., Darwin, W.D., Huestis, M.A., Cone, E.J., and Mitchell, J.M., Validity testing of the *accu*PINCH™ THC test. *J. Anal. Toxicol.*, 19, 5, 1995.
12. Kadehjian, L.J., Performance of five non-instrumented urine drug-testing devices with challenging near-cutoff specimens. *J. Anal. Toxicol.*, 25, 870, 2001.
13. Crouch, D.J., Frank, J.F., Farrell, L.J., Karsh, H.M., and Klaunig, J.E., A multiple-site laboratory evaluation of three on-site urinalysis drug-testing devices. *J. Anal. Toxicol.*, 22, 493, 1998.
14. Peace, M.R., Tarnai, L.D., and Poklis, A., Performance evaluation of four on-site drug-testing devices for detection of drugs of abuse in urine. *J. Anal. Toxicol.*, 24, 589, 2000.
15. Peace, M.R., Poklis, J.L., Tarnai, L.D., and Poklis, A., An evaluation of the OnTrak Testcup® on-site urine drug-testing device for drugs commonly encountered from emergency departments. *J Anal. Toxicol.*, 26, 500, 2002.
16. Crouch, D.J., Hersch, R.K., Cook, R.F., Frank, J.F., and Walsh, J.M., A field evaluation of five on-site drug-testing devices. *J. Anal. Toxicol.*, 26, 493, 2002.
17. Crouch, D.J., Cheever, M.L., Andrenyak, D.M., Kuntz, D.J., and Loughmiller, D.L., A comparison of ONTRAK TESTCUP, abuscreen ONTRAK, abuscreen ONLINE, and GC/MS urinalysis test results. *J. Forensic Sci.*, 43, 35, 1998.
18. Taylor, E.H., Oertli, E.H., Wolfgang, J.W., and Mueller, E., Accuracy of five on-site immunoassay drugs-of-abuse testing devices. *J. Anal. Toxicol.*, 23, 119, 1999.
19. Wu, A.H.B., Wong, S.S., Johnson, K.G., Callies, J., Shu, D.X., Dunn, W.E., and Wong, S.H.Y., Evaluation of the Triage system for emergency drugs-of-abuse testing in urine. *J. Anal. Toxicol.*, 17, 241, 1993.
20. Gronholm, M. and Lillsunde, P., A comparison between on-site immunoassay drug-testing devices and laboratory results. *Forensic Sci. Int.*, 121, 37, 2001.
21. Baker, D.P., Murphy, M.S., Shepp, P.F., Royo, V.R., Calderone, M.E., Escoto, B., and Salamone, S.J., Evaluation of the Abuscreen OnLine assay for amphetamines on the Hitachi 737: comparison with EMIT and GCMS methods. *J. Forensic Sci.*, 40, 108, 1995.
22. Fitzgerald, R.L., Ramos, J.M., Jr., Bogema, S.C., and Poklis, A., Resolution of methamphetamine stereoisomers in urine drug testing: urinary excretion of R (–)-methamphetamine following use of nasal inhalers. *J. Anal. Toxicol.*, 12, 255, 1988.
23. Cody, J.T., Important issues in testing for amphetamines, in *Handbook of Workplace Drug Testing*, Liu, R.H. and Goldberger, B.A., Eds., AACC Press, Washington, D.C., 1995, chap. 10.
24. Liu, R.H., Evaluation of commercial immunoassay kits for effective workplace drug testing, in *Handbook of Workplace Drug Testing*, Liu, R.A. and Goldberger, B.A., Eds., AACC Press, Washington, D.C., 1995.
25. ElSohly, M.A. and Jones, A.B., Drug testing in the workplace: could a positive test for one of the mandated drugs be for reasons other than illicit use of the drug? *J. Anal. Toxicol.*, 19, 450, 1995.
26. Huestis, M.A. and Cone, E.J., Drug test findings resulting from unconventional drug exposure, in *Handbook of Workplace Drug Testing*, Liu, R.A. and Goldberger, B.A., Eds., AACC Press, Washington, D.C., 1995.

27. Cone, E.J., Johnson, R.E., Darwin, W.D., Yousefnejad, D., Mell, L.D., Paul, B.D., and Mitchell, J., Passive inhalation of marijuana smoke: urinalysis and room air levels of delta-9-tetrahydrocannabinol. *J. Anal. Toxicol.,* 11, 89, 1987.

28. Hayden, J.W., Passive inhalation of marijuana smoke: a critical review. *J. Subst. Abuse,* 3, 85, 1991.

29. Siegel, R.K., ElSohly, M.A., Plowman, T., Rury, P.M., and Jones, R.T., Cocaine in herbal tea. *J. Am. Med. Assoc.,* 255, 40, 1986.

30. ElSohly, M.A., Stanford, D.F., and ElSohly, H.N., Coca tea and urinalysis for cocaine metabolites. *J. Anal. Toxicol.,* 10, 256, 1986.

31. Jackson, G.F., Saady, J.J., and Poklis, A., Urinary excretion of benzoylecgonine following ingestion of Health Inca Tea. *Forensic Sci. Int.,* 49, 57, 1991.

32. Baselt, R.C. and Chang, R., Urinary excretion of cocaine and benzoylecgonine following oral ingestion in a single subject. *J. Anal. Toxicol.,* 11, 81, 1987.

33. ElSohly, M.A. and Jones, A.B., Origin of morphine and codeine in biological specimens, in *Handbook of Workplace Drug Testing,* Liu, R.A. and Goldberger, B.A., Eds., AACC Press, Washington, D.C., 1995.

34. Changes to the testing cut-off levels for opiates for federal workplace drug testing programs. *Fed. Register,* 62, 51118, 1997.

35. Emit d.a.u. phencyclidine assay package insert; Dade-Behring, May 1997.

36. Information from Drug Abuse Warning Network: http://dawninfo.samhsa.gov/.

37. Watson, J.T., *Introduction to Mass Spectrometry,* Raven Press, New York, 1985.

38. Deutsch, D.G., Gas chromatography/mass spectrometry using table top instruments, in *Analytical Aspects of Drug Testing,* Deutsch, D.G., Ed., John Wiley & Sons, New York, 1989, chap. 4.

39. Goldberger, B.A., Huestis, M.A., and Wilkins, D.G., Commonly practiced quality control and quality assurance procedures for gas chromatography/mass spectrometry analysis in forensic urine drug-testing laboratories. *Forensic Sci. Rev.,* 9, 59, 1997.

40. The Draft Guideline on Validation of Analytical Procedures, *Fed. Register,* 62, 9315, 1996.

41. Stevenson, R., An open letter to the Food and Drug Administration, *Am. Lab.,* 28, 4, 1996.

42. Karnes, H.T., Shiu, G., and Shah, V.P., Review. Validation of bioanalytical methods. *Pharm. Res.,* 8, 421, 1991.

43. Dadgar, D., Burnett, P.E., Choc, M.G., Gallicano, K., and Hooper, J.W., Application issues in bioanalytical method validation, sample analysis and data reporting. *J. Pharm. Biomed. Anal.,* 13, 89, 1995.

44. Arnoux, P. and Morrison, R., Drug analysis of biological samples: a survey of validation approaches in chromatography in the United Kingdom pharmaceutical industry. *Xenobiotica,* 22, 757, 1992.

45. Inman, E.L., Frishmann, J.K., Jimenez, P.J., Winkel, G.D., Persinger, M.L., and Rutherford, B.S., General method validation guidelines for pharmaceutical samples. *J. Chrom. Sci.,* 25, 252, 1987.

46. Karnes, H.T. and March, C., Precision, accuracy, and data acceptance criteria in biopharmaceutical analysis. *Pharm. Res.,* 10, 1420, 1993.

47. Bonate, P.L., Concepts in calibration theory. I. Regression; *LC-GC,* 10, 310, 1992.

48. Bonate, P.L., Concepts in calibration theory. II. Regression through the origin — when should it be used? *LC-GC,* 10, 378, 1992.

49. Hornbeck, C.L., Carrig, J.E., and Czarny, R.J., Detection of a GC/MS artifact peak as methamphetamine. *J. Anal. Toxicol.,* 17, 257, 1993.

50. Fenton, J., Mummert, J., and Childers, M., Hydromorphone and hydrocodone interference in GC/MS assays for codeine and morphine. *J. Anal. Toxicol.,* 18, 159, 1994.

51. ElSohly, M.A., Stanford, D.F., Sherman, D., Shah, H., Bernot, D., and Turner, C.E., A procedure for eliminating interferences from ephedrine and related compounds in the GC/MS analysis of amphetamine and methamphetamine. *J. Anal. Toxicol.,* 16, 109, 1992.

52. Paul, B.D., Past, M.R., McKinley, R.M., Foreman, J.D., McWhorter, L.K., and Snyder, J.J., Amphetamine as an artifact of methamphetamine during periodate degradation of interfering ephedrine, pseudoephedrine, and phenylpropanolamine: an improved procedure for accurate quantitation of amphetamines in urine. *J. Anal. Toxicol.,* 18, 331, 1994.

53. Kaplan, L.A. and Pesce, A.J., *Clinical Chemistry: Theory, Analysis, and Correlation,* C.V. Mosby, St. Louis, MO, 1984.

54. Smith, R.V. and Stewart, J.T., *Textbook of Biopharmaceutic Analysis,* Lea & Febiger, Philadelphia, 1981.

55. Wu Chen, N.B., Cody, J.T., Garriott, J.C., Foltz, R.L., Peat, M.A., and Schaffer, M.I., Report of the 1988 Ad Hoc Committee on Forensic GC/MS: Recommended guidelines for forensic GC/MS procedures in toxicology laboratories associated with offices of medical examiners and/or coroners. *J. Forensic Sci.,* 35, 236, 1990.

56. Guidance Document for Laboratories and Inspectors, U.S. Department of Health and Human Services, National Laboratory Certification Program, Revised August 29, 1994.

57. Kaplan, A. and Szabo, L., *Clinical Chemistry: Interpretation and Techniques,* 2nd ed., Lea & Febiger, Philadelphia, 1983.

58. Moffat, A.C., Jackson, J.V., Moss, M.S., and Widdop, B., Eds., *Clarke's Isolation and Identification of Drugs,* 2nd ed., Pharmaceutical Press, London, 1986.

59. Poole, C.F. and Shuette, S., *Contemporary Practice of Chromatography,* Elsevier, New York, 1984.

60. Grob, K., Grob, G., and Grob, K., Jr., Testing capillary gas chromatographic columns. *J. Chromatogr.,* 219, 13, 1981.

61. McNair, H.M., Method development in gas chromatography. *LC-GC,* 11, 94, 1993.

62. Cembrowski, G.S. and Carey, R.N., *Laboratory Quality Management.* ASCP Press, Chicago, IL, 1989.

63. Westgard, J.O., Better quality control through microcomputers. *Diagn. Med.,* 61, 741, 1982.

64. Westgard, J.O., Barry, P.L., Hunt, M.R., and Groth, T., A multi-rule Shewart chart for quality control in clinical chemistry. *Clin. Chem.,* 27, 493, 1981.

65. Westgard, J.O. and Barry, P.L., *Cost-Effective Quality Control: Managing the Quality and Productivity of Analytical Processes.* AACC Press, Washington, D.C., 1986.

66. Shewhart, W.A., *Statistical Method from the Viewpoint of Quality Control.* Dover, New York, 1986.

67. Garfield, F.M., *Quality Assurance Principles for Analytical Laboratories,* AOAC International, Gaithersburg, MD, 1991.

68. Cairns, T., Siegmund, E.G., and Stamp, J.J., Evolving criteria for confirmation of trace level residues in food and drugs by mass spectrometry. Part I. *Mass Spectrom. Rev.,* 8, 93, 1989.

69. Cairns, T., Siegmund, E.G., and Stamp, J.J., Evolving criteria for confirmation of trace level residues in food and drugs by mass spectrometry. Part II. *Mass Spectrom. Rev.,* 8, 127, 1989.

70. Sphon, J.A., Use of mass spectrometry for confirmation of animal drug residues. *J. Assoc. Off. Anal. Chem.,* 61, 1247, 1978.

71. Garland, W.A. and Barbalas, M.P., Applications to analytic chemistry: an evaluation of stable isotopes in mass spectral drug assays. *J. Clin. Pharm.,* 26, 412, 1986.

72. Lawson, A.M., Gaskell, S.J., and Hjelm, M., Methodological aspects on quantitative mass spectrometry used for accuracy control in clinical chemistry. International Federation of Clinical Chemistry (IFCC), Office for Reference Methods and Materials (ORMM). *J. Clin. Chem. Clin. Biochem.,* 23, 433, 1985.

73. Knapp, D.R., *Handbook of Analytical Derivatization Reactions,* John Wiley & Sons, New York, 1979.

74. Blau, K. and King, G.S., Eds., *Handbook of Derivatives for Chromatography,* Heyden & Son, London, 1978.

75. Blau, K. and Halket, J.M., Eds., *Handbook of Derivatives for Chromatography,* 2nd ed., John Wiley & Sons, New York, 1993.

76. Knapp, D.R., Chemical derivatization for mass spectrometry. *Methods Enzymol.,* 193, 314, 1990.

77. Moore, J.M., The application of chemical derivatization in forensic drug chemistry for GC and HPLC methods of analysis. *Forensic Sci. Rev.,* 2, 80, 1990.

78. Seno, H., Suzuki, O., and Kumazawa, T., Rapid isolation with Sep-Pak C18 cartridges and wide-bore capillary gas chromatography of benzophenones, the acid-hydrolysis products of benzodiazepines. *J. Anal. Toxicol.,* 15, 21, 1991.

79. Foltz, R.L., Fentiman, A.F., and Foltz, R.B., GC/MS Analysis for Abused Drugs in Body Fluids, NIDA Research Monograph 32, August 1980.

80. Millard, B.J., *Quantitative Mass Spectrometry,* Heyden, London, 1978.

81. Liu, R.H., Baugh, D.L., Allen, E.E., Salud, S.C., Fentress, J.G., Chadha, H., and Walia, A.S., Isotopic analogue as the international standard for quantitative determination of benzoylecgonine: concerns with isotopic purity and concentration level. *J. Forensic Sci.,* 34, 986, 1989.

82. Liu, R.H., McKeehan, A.M., Edwards, C., Foster, G., Bensley, W.D., Langner, J.G., and Walia, A.S., Improved gas chromatography/mass spectrometry analysis of barbiturates in urine using centrifuge-based solid-phase extraction, methylation with d5-pentobarbital as internal standard. *J. Forensic Sci.,* 39, 1501, 1994.

83. Liu, R.H., Foster, G., Cone, E.J., and Kumar, S.D., Selecting an appropriate isotopic internal standard for gas chromatography/mass spectrometry analysis of drugs of abuse — pentobarbital example. *J. Forensic Sci.*, 40, 983, 1995.

84. Thurman, E.M., Pedersen, M.J., Stout, R.L., and Martin, T., Distinguishing sympathomimetic amines from amphetamine and methamphetamine in urine by gas chromatography/mass spectrometry. *J. Anal. Toxicol.*, 16, 19, 1992.

85. Hughes, R.O., Bronner, W.E., and Smith, M.L., Detection of amphetamine and methamphetamine in urine by gas chromatography/mass spectrometry following derivatization with (–)-methyl chloroformate. *J. Anal. Toxicol.*, 15, 256, 1991.

86. Melgar, R. and Kelly, R.C., A novel GC/MS derivatization method for amphetamines. *J. Anal. Toxicol.*, 17, 399, 1993.

87. Mule, S.J. and Casella, G.A., Confirmation and quantitation of barbiturates in human urine by gas chromatography/mass spectrometry. *J. Anal. Toxicol.*, 13, 13, 1989.

88. Dickson, P.H., Markus, W., McKernan, J., and Nipper, H.C., Urinalysis of alpha-hydroxyalprazolam, alpha-hydroxytriazolam, and other benzodiazepine compounds by GC/EIMS. *J. Anal. Toxicol.*, 16, 67, 1992.

89. Fitzgerald, R.L., Rexin, D.A., and Herold, D.A., Benzodiazepine analysis by negative ion chemical ionization gas chromatography/mass spectrometry. *J. Anal. Toxicol.*, 17, 342, 1993.

90. Black, D.A., Clark, G.D., Haver, V.M., Garbin, J.A., and Saxon, A.J., Analysis of urinary benzodiazepines using solid-phase extraction and gas chromatography-mass spectrometry. *J. Anal. Toxicol.*, 18, 185, 1994.

91. Kemp, P.M., Abukhalaf, I.K., Manno, J.E., Manno, B.R., Alford, D.D., and Abusada, G.A., Cannabinoids in humans. I. Analysis of delta-9-tetrahydrocannabinol and six metabolites in plasma and urine using GC-MS. *J. Anal. Toxicol.*, 19, 285, 1995.

92. Kemp, P.M., Abukhalaf, I.K., Manno, J.E., Manno, B.R., Alford, D.D., Mcwilliams, M.E., Nixon, F.E., Fitzgerald, M.J., Reeves, R.R., and Wood, M.J., Cannabinoids in humans. II. The influence of three methods of hydrolysis on the concentration of THC and two metabolites in urine. *J. Anal. Toxicol.*, 19, 292, 1995.

93. Joern, W.A., Detection of past and recurrent marijuana use by a modified GC/MS procedure. *J. Anal. Toxicol.*, 11, 49, 1987.

94. Paul, B.D., Mell, L.D., Mitchell, J.M., and McKinley, R.M., Detection and quantitation of urinary 11-nor-delta-9-tetrahydrocannabinol-9-carboxylic acid, a metabolite of tetrahydrocannabinol by capillary gas chromatography and electron impact mass fragmentography. *J. Anal. Toxicol.*, 11, 1, 1987.

95. Cone, E.J., Hillsgrove, M., and Darwin, W.D., Simultaneous measurement of cocaine, cocaethylene, their metabolites and "crack" pyrolysis products by gas chromatography-mass spectrometry. *Clin. Chem.*, 40, 1299, 1994.

96. Mule, S.J. and Casella, G.A., Confirmation and quantitation of cocaine, benzoylecgonine, ecgonine methyl ester in human urine by GC/MS. *J. Anal. Toxicol.*, 12, 153, 1988.

97. Goldberger, B.A., Darwin, W.D., Grant, T.M., Allen, A.C., Caplan, Y.H., and Cone, E.J., Measurement of heroin and its metabolites by isotope-dilution electron-impact mass spectrometry. *Clin. Chem.*, 39, 670, 1993.

98. Mitchell, J.M., Paul, B.D., Welch, P., and Cone, E.J., Forensic drug testing for opiates. II. Metabolism and excretion rate of morphine in humans after morphine administration. *J. Anal. Toxicol.*, 15, 49, 1991.

99. Cone, E.J., Welch, P., Mitchell, J.M., and Paul, B.D., Forensic drug testing of opiates. I. Detection of 6-acetylmorphine in urine as an indicator of recent heroin exposure. Drug and assay considerations and detection times. *J. Anal. Toxicol.*, 15, 1, 1991.

100. ElSohly, H.N., ElSohly, M.A., and Stanford, D.F., Poppy seed ingestion and opiate urinalysis: a closer look. *J. Anal. Toxicol.*, 14, 308, 1990.

CHAPTER 7

Analytical Approaches for Drugs in Biological Matrices Other Than Urine

Pascal Kintz, Pharm.D., Ph.D., Marion Villain, M.S., and Vincent Cirimele, Ph.D.
ChemTox Laboratory, Illkirch, France

CONTENTS

It is generally accepted that chemical testing of biological fluids is the only objective means of diagnosis of drug use. The presence of a drug analyte in a biological specimen can be used as evidence of exposure. The standard procedure in drug testing is the immunoassay screening, followed by the gas chromatographic/mass spectrometric (GC/MS) confirmation conducted on a urine sample.[1,2] More recently, a variety of body specimens other than urine, such as oral fluid, sweat, or hair, have been proposed to document drug exposure. It appears that the value of alternative specimen analysis for the identification of drug users is steadily gaining recognition. This can be seen from its growing use in pre-employment screening, in forensic sciences, and in clinical applications.[3-5]

The advantages of these samples over traditional media like urine and blood are that collection is almost non-invasive, relatively easy to perform, and may be achieved under close supervision to prevent adulteration or substitution of the sample. Tools for the detection of drugs in alternative specimens utilize traditional technology, although some limitations are imposed that require special attention:

1. The specimen volume or mass is often small.
2. The target analytes are different from urine.
3. The analyte concentration is lower than in urine.
4. The sample preparation for drug analysis can be more difficult.

Although remarkable advances in sensitive analytical techniques enable drug confirmation in oral fluid, sweat, and hair today,[6] limited progress has been made in the development of commercial collection devices, on-site and laboratory-based commercial screening assays, quality control materials, proficiency-testing programs, and regulatory guidelines for utilization of alternative matrices in drug testing.

REFERENCES

1. Substance Abuse and Mental Health Administration, Mandatory guidelines for Federal workplace drug testing programs, *Fed. Regis.*, 53, 11970, 1988.
2. de la Torre, R. et al., Recommendations for the reliable detection of illicit drugs in urine in the European Union with special attention to the workplace, *Ann. Clin. Biochem.*, 34, 339, 1997.
3. Caplan, Y.H. and Goldberger, B.A., Alternative specimens for workplace drug testing, *J. Anal. Toxicol.*, 25, 396, 2001.
4. UNDCP, *Guidelines for Testing under International Control in Hair, Sweat and Saliva*, United Nations, New York, 1998.
5. Cone, E., Legal, workplace, and treatment drug testing with alternate biological matrices on a global scale, *Forensic Sci. Int.*, 21, 7, 2001.
6. Kintz, P. and Samyn, N., Unconventional samples and alternative matrices. In *Handbook of Analytical Separations: Forensic Science*, M. Bogusz, Ed., Elsevier, Amsterdam, 2000, 459.

7.1 HAIR

In the 1960s and 1970s, hair analysis was used to evaluate exposure to toxic heavy metals, such as arsenic, lead, or mercury. This was achieved using atomic absorption spectroscopy that allowed detection in the nanogram range. At that time, examination of hair for organic substances, especially drugs, was not possible because analytical methods were not sensitive enough. Examination by means of drugs marked with radioactive isotopes, however, established that these substances can move from blood to hair and are deposited there. Ten years after these first investigations, it was possible to demonstrate the presence of various organic drugs in hair by means of radioimmunoassay (RIA). In 1979, Baumgartner and colleagues[1] published the first report on the detection of morphine in the hair of heroin abusers using RIA. They found that differences in the concentration of morphine along the hair shaft correlated with the time of drug use. Today, GC/MS is the method of choice for hair analysis and this technology is routinely used to document repetitive drug exposure in forensic science, traffic medicine, occupational medicine, clinical toxicology, and more recently in sports.

The major practical advantage of hair testing compared to urine or blood testing for drugs is that it has a larger surveillance window (weeks to months, depending on the length of the hair shaft, against 2 to 4 days for most drugs). For practical purposes, the two tests complement each other. Urinalysis and blood analysis provide short-term information of an individual's drug use, whereas long-term histories are accessible through hair analysis. While analysis of urine and blood specimens often cannot distinguish between chronic use and single exposure, hair analysis can offer the distinction.

7.1.1 Hair Collection

Collection procedures for hair analysis for drugs have not been standardized. In most published studies, the samples are obtained from random locations on the scalp. Hair is best collected from the area at the back of the head, called the *vertex posterior*. Compared with other areas of the head, this area has less variability in the hair growth rate, the number of hairs in the growing phase is more constant, and the hair is less subject to age- and sex-related influences. Hair strands are cut as close as possible from the scalp, and their location on the scalp must be noted. Once collected, hair samples may be stored at ambient temperature in aluminum foil, an envelope, or a plastic tube. The sample size taken varies considerably among laboratories and depends on the drug to be analyzed and the test methodology. For example, when fentanyl or buprenorphine are investigated, a 100-mg sample is recommended. Sample sizes reported in the literature range from a single hair to 200 mg, cut as close to the scalp as possible. When sectional analysis is performed, the hair is cut into segments of about 1, 2, or 3 cm, which corresponds to about 1, 2, or 3 months' growth.

7.1.2 Decontamination Procedures

Contaminants of hair would be a problem if they were drugs of abuse or their metabolites or if they interfered with the analysis and interpretation of the test results. It is unlikely that anyone would intentionally or accidentally apply anything to their hair that would contain a drug of abuse. The most crucial issue facing hair analysis is the avoidance of technical and evidentiary false positives. Technical false positives are caused by errors in the collection, processing, and analysis of specimens, while evidentiary false positives are caused by passive exposure to the drug. Approaches for preventing evidentiary false positives due to external contamination of the hair specimens have been described.[2]

Most but not all laboratories use a wash step; however, there is no consensus or uniformity in the washing procedures. Among the agents used in washing are detergents such as shampoo, surgical scrubbing solutions, surfactants such as 0.1% sodium dodecylsulfate, phosphate buffer, or organic solvents such as acetone, diethyl ether, methanol, ethanol, dichloromethane, hexane, or pentane of various volumes for various contact times. Generally, a single washing step is used, although a second identical wash is sometimes performed. If external contamination is found by analyzing the wash solution, the washout kinetics of repeated washing can demonstrate that contamination is rapidly removed. According to Baumgartner and Hill,[2] the concentration of drug in the hair after washing should exceed the concentration in the last wash by at least ten times. It has also been proposed that hair should be washed three times with phosphate prior to analysis to remove any possible external contamination and that the total concentration of any drug present in the three phosphate washes should be greater than 3.9 times the concentration in the last wash.

Detection of drug metabolites in hair, whose presence could not be explained by hydrolysis or environmental exposure, would unequivocally establish that internal drug exposure had occurred.[3] Cocaethylene and nor-cocaine would appear to meet these criteria, as these metabolites are formed only when cocaine is metabolized. Because these metabolites are not found in illicit cocaine samples, they would not be present in hair as a result of environmental contamination, and thus their presence in hair could be considered a marker of cocaine exposure. This procedure can be extended to other drugs. However, there is still a great controversy about the potential risk of external contamination, particularly for crack, cannabis, and heroin when smoked, as several authors have demonstrated that it is not possible to fully remove the drugs.[4,5] In conclusion, although it is highly recommended to include a decontamination step, there is no consensus on which procedure performs best and each laboratory must validate its own technique.

7.1.3   Drug Solubilization

To determine the amount of a drug remaining in hair after washing, it is necessary to solubilize the drugs in the hair. Solubilization must be such that the analytes are not altered or lost. Care is necessary to prevent the conversion of cocaine to benzoylecgonine or 6-monoacetylmorphine to morphine, for example.

The hair sample can be pulverized in a ball-mill prior to testing, cut into segments, or the entire hair dissolved. The preparation techniques are generally based on one of the following procedures:

- Incubation in an aqueous buffer and analysis with immunological techniques, mostly RIA
- Incubation in an acidic or basic solution followed by liquid–liquid extraction or solid-phase extraction and analysis with chromatographic techniques, mostly GC/MS
- Incubation in an organic solvent (generally methanol with or without hydrochloric acid), liquid extraction or solid-phase extraction, and analysis with chromatographic techniques, mostly GC/MS
- Digestion in an enzymatic solution, liquid extraction or solid-phase extraction, and analysis with chromatographic techniques, mostly GC/MS

7.1.4   Drug Analysis

The first publication dealing with the analysis of morphine in hair for determining the history of opiate abuse reported the use of RIA.[1] This paper was followed by a great number of procedures that mostly used RIA and/or GC/MS. Chromatographic procedures are a powerful tool for the identification and quantification of drugs in hair, owing to their separation ability and their detection sensitivity and specificity, particularly when coupled with MS. Proposed cutoff concentrations and expected concentrations for drugs of abuse in hair are presented in Table 7.1.

Radioimmunoassay is the most common screening test for hair. Kits, generally designed for urine, can be used without any modification, at pH values above 7. Calibration curves are obtained either from the controlled urines in the kit or from extracts of drug-free hair samples spiked with the drugs. Duplicate determinations are recommended. The RIA results should be confirmed by GC/MS. In the absence of a second independent method, RIA detection must be interpreted with caution. However, even the high sensitivity of GC/MS is sometimes not sufficient to detect drugs, especially when starting with a small quantity of hair. For these reasons, it may be necessary to carry out immunological analysis of drugs in hair using RIA reagents that are specific for the selective estimation of a drug like fentanyl, lysergide (LSD), or buprenorphine.

Table 7.1   Proposed Cutoff Concentrations (when tested by GC/MS) and Expected Concentrations for Drugs of Abuse in Hair

Drug	GC/MS Cutoff Concentration	Expected Concentrations
Heroin	0.5 ng/mg of 6-acetylmorphine	0.5–100 ng/mg, in most cases <15 ng/mg
Cocaine	0.5 ng/mg of cocaine	0.5–100 ng/mg, in most cases <50 ng/mg, in crack abusers >300 ng/mg is possible
Amphetamine, MDMA	0.5 ng/mg for both drugs	0.5–50.0 ng/mg
Cannabis	0.05 ng/mg for THC 0.5 pg/mg for THC-COOH	THC: 0.05–10 ng/mg, in most cases <3 ng/mg THC-COOH: 0.5–50 pg/mg, in most cases <5 pg/mg

Table 7.2   Screening Procedures for the Detection of Controlled Drugs in Hair (see Reference 9)

Method	Kauert	Moeller	Kintz
Analytes	Heroin, 6-MAM, dihydrocodeine, codeine, methadone, THC, cocaine, amphetamine, MDMA, MDEA, MDA	6-MAM, dihydrocodeine, codeine, methadone, THC, cocaine, amphetamine, MDMA, MDEA, MDA	6-MAM, codeine, methadone, cocaine, amphetamine, MDMA, MDEA, MDA, most pharmaceuticals
Decontamination step	Ultrasonic 5 min each 5 ml H_2O 5 ml acetone 5 ml petroleum ether	20 ml H_2O (2×) 20 ml acetone	5 ml dichloromethane (2 × 5 min)
Homogenization	100 mg hair cut into small sections in a 30-ml vial	Ball mill	Ball mill
Extraction	4 ml methanol, ultrasonic 5 h, 50°C	20–30 mg powdered hair, 2 ml acetate buffer + β-glucuronidase/aryl-sulfatase, 90 min/40°C	50 mg powdered hair, 1 ml 0.1 N HCl, 16 h/56°C
Clean-up	None	$NaHCO_3$; SPE (C18), elution with 2 ml acetone/dichloromethane (3:1)	$(NH_4)_2HPO_4$; extraction 10 ml chloroform/2-propanol/n-heptane (50:17:33); organic phase purified with 0.2 N HCl; HCl phase to pH 8.4; re-extraction with $CHCl_3$
Derivatization	Propionic acid anhydride	1000 μl PFPA/75 μl PF-n-propanol; 30 min/60°C; N_2/60°C; 50 μl ethylacetate	40 μl BSTFA + 1% TMCS or HFBA; 20 min/70°C

Chromatographic methods have been used as screening and confirming tests. Moreover, they allow quantification of the drugs and drug metabolites. Some classic GC/MS procedures are given in detail in Table 7.2. There are very few papers that present data using high-performance liquid chromatography/MS (HPLC/MS), except for buprenorphine, diuretics, and corticosteroids.

7.1.5   Drug Identification

7.1.5.1   Opiates

As heroin samples always contain codeine, codeine is also detected in cases of heroin abuse. Morphine is a metabolite of codeine and can be detected when codeine is abused. The quantification of both drugs was proposed to differentiate between codeine and heroin abuse.[6] If the morphine concentration is clearly higher than the codeine concentration in the hair sample, heroin or morphine abuse is highly probable. If the codeine concentration is higher than the morphine level, then it may be assumed that codeine has been ingested. However, the discrimination of heroin users from individuals exposed to other sources of morphine alkaloids can be achieved by identifying directly heroin or 6-monoacetylmorphine.[7,8] In this case, alkaline hydrolysis must not be used in order to avoid hydrolysis. In most samples, it was demonstrated that 6-monoacetylmorphine concentrations exceeded those of morphine, which is a less lipophilic compound. Other opioids have been detected, including dihydrocodeine, pholcodine, ethylmorphine, dextromoramide, methadone, fentanyl, sufentanyl, pentazocine, zipeprol, and buprenorphine.

7.1.5.2   Cocaine

Procedures for the detection of cocaine have been published in several papers.[9,10] There is considerable variety in the workup and derivatization conditions. In most cases, cocaine is found in higher concentrations than benzoylecgonine and methylecgonine.[11,12] The determination of the pyrolysis product of cocaine, anhydroecgonine methylester (AEME), is helpful to distinguish cocaine from crack users. Kintz found AEME in the range 0.2 to 2.4 ng/mg for seven crack users.[13] An important study with controlled doses of cocaine-d_5 was published in 1996.[14] The deuterium-labeled cocaine was administered intravenously and/or intranasally in doses of 0.6 to 4.2 mg/kg under controlled conditions. A single dose could be detected for 2 to 6 months; the minimum detectable dose appeared to be between 22 and 35 mg.

7.1.5.3 *Amphetamines*

Papers dealing with amphetamine and methamphetamine have been published.[15] The workup (liquid–liquid extraction after acid or alkaline hydrolysis) and derivatization procedures (trifluoro-acetic anhydride, TFA) are similar in most of the publications. After methamphetamine intake, its major metabolite, amphetamine, can be detected in hair samples and the differentiation between methamphetamine and amphetamine intake can be achieved by reviewing the ratio between the drugs. In 1992, methylenedioxy-*N*-methamphetamine (MDMA or ecstasy) was first detected in the hair of an abuser at the concentration of 0.6 ng/mg.[16] In Europe, MDMA is one of the most frequently identified drugs. Since the first screening procedure in 1995, different methods have been published.[17,18] These procedures permit the determination of amphetamine, methylamphet-amine, methylenedioxyamphetamine, methylenedioxymethamphetamine, methylenedioxyetham-phetamine, and *N*-methylbenzodioxazolybutanamine (MBDB) by electron impact GC/MS.

7.1.5.4 *Cannabis*

In 1995 the first results on cannabis in hair by using GC/MS and the determination in the same run of Δ^9-tetrahydrocannabinol (THC) and its major metabolite 11-nor-Δ^9-THC carboxylic acid (THC-COOH) were reported.[19] The measured concentrations were low, particularly in comparison with other drugs. Some authors suggested the use of negative chemical ionization[20] to target the drugs or the application of tandem mass spectrometry.[21] More recently, a simpler method was proposed,[22] based on the simultaneous identification of cannabinol (CBN), cannabidiol (CBD), and THC. This procedure is a screening method that is rapid, economic, and does not require deriva-tization prior to analysis. To avoid potential external contamination (as THC, CBD, and CBN are present in smoke), the endogenous metabolite THC-COOH should be looked for to confirm drug use. The concentrations measured are very low, particularly for THC-COOH (pg/mg range).

7.1.6 Place of Hair

Although there are still controversies on how to interpret the results, particularly concerning external contamination, cosmetic treatments, genetic considerations, and drug incorporation, pure analytical work in hair analysis has reached a sort of plateau, having solved almost all the analytical problems. Although GC/MS is the method of choice in practice, GC/MS/MS or LC/MS are today used in several laboratories, even for routine cases, particularly to target low-dosage compounds, such as THC-COOH, fentanyl, flunitrazepam, or buprenorphine. Electrophoretic/electrokinetic ana-lytical strategies, chiral separation, or application of ion mobility spectrometry constitute the latest new developments of the analytical tools reported to document the presence of drugs in hair. Quality assurance is a major issue of drug testing in hair. Since 1990, the National Institute of Standards and Technology (Gaithersburg, MD) has developed interlaboratory comparisons, recently followed by the Society of Hair Testing (Strasbourg, France). At the moment, some laboratory guidelines concerning hair analysis are under preparation, both by the U.S. Substance Abuse Mental Health Services Administration (SAMHSA) Mandatory Guidelines and the Society of Hair Testing; how-ever, final publication has not yet materialized.

By providing information on exposure to drugs over time, hair analysis may be useful in verifying self-reported histories of drug use in any situation in which a history of past rather than recent drug use is desired, as in pre-employment and employee drug testing. In addition, hair analysis may be especially useful when a history of drug use is difficult or impossible to obtain, such as from psychiatric patients. During control tests of hair fragments, a drug addict is not able to hide the fact that drugs have been used, whereas intermittent drug use may be difficult to detect if urine or blood tests alone are undertaken, even when the tests are repeated. There are essentially three types of problems with urine drug testing: false positives when not confirmed with GC/MS, the embarrassment associated with observed urine collection, and evasive maneuvers, including adulteration. These problems can be greatly mitigated or eliminated through hair analysis. It is always possible to obtain a fresh, identical hair sample if there is any claim of a specimen mix-up or breach in the chain of custody. This makes hair analysis essentially fail-safe, in contrast to urinalysis, since an identical urine specimen cannot be obtained at a later date.

Another potential use of hair analysis is to verify accidental or unintentional ingestion of drinks or food that has been laced with drugs. In case of a single use, the hair will not test positive. For example, ingestion of poppy seeds appears to be sufficient for the creation of a positive urine result, while ingestion of up to 30 g of poppy seeds did not result in a positive hair identification (Sachs, personal communication, 1994). The greatest use of hair analysis, however, may be in identifying false negatives, since neither abstaining from a drug for a few days nor trying to "beat the test" by diluting urine will alter the concentration in hair. Urine does not indicate the frequency of drug intake in subjects who might deliberately abstain for several days before drug screenings. While analysis of urine specimens cannot distinguish between chronic use and single exposure, hair analysis offers the potential to make this distinction.

REFERENCES

1. Baumgartner, A.M. et al., Radioimmunoassay of hair for determining opiate-abuse histories, *J. Nuclear Med.*, 20, 748, 1979.
2. Baumgartner, W.A. and Hill, V.A., Hair analysis for drugs of abuse: decontamination issues. In *Recent Developments in Therapeutic Drug Monitoring and Clinical Toxicology,* Sunshine, I., Ed., Marcel Dekker, New York, 1992, 577.
3. Cone, E.J. et al., Testing human hair for drugs of abuse. II. Identification of unique cocaine metabolites in hair of drug abusers and evaluation of decontamination procedures, *J. Anal. Toxicol.*, 15, 250, 1991.
4. Blank, D.L. and Kidwell, D.A., Decontamination procedures for drugs of abuse in hair: are they sufficient? *Forensic Sci. Int.*, 70, 13, 1995.
5. Romano, G. et al., Determination of drugs of abuse in hair: evaluation of external heroin contamination and risk of false positives, *Forensic Sci. Int.*, 131, 98, 2003.
6. Sachs, H. and Arnold, W., Results of comparative determination of morphine in human hair using RIA and GC/MS, *J. Clin. Chem. Clin. Biochem.*, 27, 873, 1989.
7. Sachs, H. and Uhl, M., Opiat-Nachweis in Haar-Extrakten mit Hilfe von GC/MS/MS und Supercritical Fluid Extraction, *Toxichem. Krimtech.*, 59, 114, 1992.
8. Nakahara, Y. et al., Hair analysis for drugs of abuse. IV. Determination of total morphine and confirmation of 6-acetylmorphine in monkey and human hair by GC/MS, *Arch. Toxicol.*, 66, 669, 1992.
9. Sachs, H. and Kintz, P., Testing for drugs in hair. Critical review of chromatographic procedures since 1992, *J. Chromatogr. B,* 713, 147, 1998.
10. Moeller, M.R., Drug detection in hair by chromatographic procedures, *J. Chromatogr.*, 580, 125, 1992.
11. Henderson, G.L., Harkey, M.R., and Zhou, C., Cocaine and metabolite concentrations in hair of South American coca chewers, *J. Anal. Toxicol.*, 16, 199, 1992.
12. Moeller, M.R., Fey, P., and Rimbach, S., Identification and quantitation of cocaine and its metabolites, benzoylecgonine and ecgonine methylester in hair of Bolivian coca chewers by GC/MS, *J. Anal. Toxicol.*, 16, 291, 1992.
13. Kintz, P. et al., Testing human hair and urine for anhydroecgonine methylester, a pyrolysis product of cocaine, *J. Anal. Toxicol.*, 19, 479, 1995.
14. Henderson, G.L. et al., Incorporation of isotopically labeled cocaine and metabolites into human hair. I. Dose–response relationships, *J. Anal. Toxicol.*, 20, 1, 1996.

15. Nakahara, Y. et al., Hair analysis for drug abuse. II. Hair analysis for monitoring of methamphetamine abuse by isotope dilution GC/MS, *Forensic Sci. Int.,* 46, 243, 1990.
16. Moeller, M.R., Maurer, H.H., and Roesler, M., MDMA in blood, urine and hair: a forensic case. In *Proceedings of the 30th Meeting of TIAFT,* Nagata, T., Ed., Yoyodo Printing Kaisha, Fukuoka, 1992, 347.
17. Kintz, P. et al., Simultaneous determination of amphetamine, methamphetamine, MDA and MDMA in human hair by GC/MS, *J. Chromatogr. B,* 670, 162, 1995.
18. Rothe, M. et al., Hair concentrations and self-reported abuse history of 20 amphetamine and ecstasy users, *Forensic Sci. Int.,* 89, 111, 1997.
19. Cirimele, V., Kintz, P., and Mangin, P., Testing human hair for cannabis, *Forensic Sci. Int.,* 70, 175, 1995.
20. Kintz, P., Cirimele, V., and Mangin, P., Testing human hair for cannabis. II. Identification of THC-COOH by GC/MS/NCI as an unique proof, *J. Forensic Sci.,* 40, 619, 1995.
21. Uhl, M., Determination of drugs in hair using GC/MS/MS, *Forensic Sci. Int.,* 84, 281, 1997.
22. Cirimele, V. et al., Testing human hair for cannabis. III. Rapid screening procedure for the simultaneous identification of THC, cannabinol and cannabidiol, *J. Anal. Toxicol.,* 20, 13, 1996.

7.2 ORAL FLUID

Oral fluid has been increasingly used as an analytical tool in pharmacokinetic studies, thera-peutic drug monitoring, and the detection of illicit drugs. More recently, particular interest in oral fluid has been expressed by law enforcement agencies for roadside testing of intoxicated drivers or in occupational medicine. The presence of metabolites of drugs in urine of potentially impaired drivers can be interpreted as evidence of relatively recent exposure, except for cannabis. However, this does not necessarily mean that the subject was under the influence at the time of sampling. It has been claimed that the concentrations of many drugs in oral fluid correlate well with blood concentrations, which suggests that qualitative measurements in saliva may be a valuable technique to determine the current degree of exposure to a definite drug at the time of sampling.

7.2.1 Oral Fluid Sampling

Several methods have been described for the collection of mixed saliva. Based on a recommen-dation of a working group of the Drug Testing Advisory Board in the U.S., the term *oral fluid* is preferred. Saliva stands for the secretion of the salivary glands, while the testing fluid also contains mucosal transsudate and crevicular fluid.

Oral fluid is usually collected by spitting into a collection vial or wiping the oral cavity with a swab. Many different stimuli will cause salivation. Spitting itself is usually a sufficient stimulus to elicit a flow of about 0.5 ml/min. However, there are a number of limitations: in an on-site situation "dry mouth" was often reported;[1] moreover, the salivary flow is decreased after intake of certain drugs, e.g., some tricyclic antidepressants and amphetamines. Individuals often collect more froth than actual liquid providing a viscous sample with a small sample size, which com-plicates the analysis. Many researchers have found it advantageous to stimulate salivation by placing sour candy or citric acid crystals in the mouth prior to collection, or chewing an inert material like Teflon. Substances like Parafilm® should be avoided because they may absorb highly lipophilic drugs. The individual should allow saliva to accumulate in the mouth and then expec-torate into a suitable container. Some special devices have been designed to facilitate the sampling of saliva. The use of a dental cotton roll to collect saliva has been improved and is nowadays available as the Salivette® (Sarstedt, Germany). Other commercialized devices include Accu-sorb®, SmartClip®, and Intercept®.

The effect of collection methods on drug concentrations in oral fluid is not well described in the scientific literature. However, caution has to be taken. As the oral fluid flow rate increases, the concentration of bicarbonate ions increases. This may alter the saliva/plasma ratio in a pH-dependent manner, especially for weakly basic drugs with a pK_a approaching the oral fluid pH, e.g., cocaine and opioids.[2] Kato et al.[3] observed that the concentrations of cocaine, benzoylecgonine, and ecgonine methylester were substantially greater in nonstimulated mixed saliva than after stimulation with citric acid. A controlled clinical study designed to determine the effects of selected collection protocols on oral fluid codeine concentrations was published recently.[4] Concentrations at different time points averaged 3.6 times higher after spitting without stimulation (= control method) than with acidic stimulation. The control method showed 1.3 to 2.0 times higher concentrations than concentrations in specimens collected by chewing a sugarless gum or using the Salivette or the Finger Collector containing Accu-Sorb® (Avitar Technologies Inc., Canton, MA, U.S.A.). A second issue influencing the use of oral fluid as a drug testing matrix is the contamination of the buccal cavity depending on the route of administration, e.g., for several hours after smoking or sniffing certain drugs. After smoking a cannabis cigarette, the cannabinoids in the smoke are sequestered in the mouth. THC concentrations obtained after direct extraction of a Salivette (positioned between gum and cheek for 1 to 2 min) exceeded the concentrations of THC in samples obtained by spitting by a factor 10 to 100.

7.2.2   Screening Tests for Oral Fluid

Before oral fluid can be used for rapid screening and especially for on-site testing, the following criteria should be met: (1) a fast, simple, and validated sampling procedure; (2) a test that needs a small sample volume (100 µl); (3) an antibody targeted to the parent molecules rather than the metabolites; (4) a sensitivity adapted to the expected concentrations in oral fluid; (5) screening cutoff values that meet the requirements for high sensitivity and specificity; (6) an electronic reader.

Some prototypes of on-site tests have been investigated during the course of the ROSITA project.[5] Unfortunately, none of the present devices is satisfactory in terms of sensitivity and reliability. However, many development efforts are under way and the involved companies are improving the ease of use and the accuracy of their tests.

The results of the first large-scale database on oral fluid testing in private industry were published in 2002.[6] About 77,000 specimens were screened by the Intercept immunoassay at manufacturer's recommended cutoff values for five drug classes (cannabis, opiates, amphetamines, cocaine, and phencyclidine). Presumptive positive specimens were confirmed by GC/MS/MS.

The screening cutoffs will depend on the specificity of the antibodies, and the presence of other cross-reacting metabolites in oral fluid. Quite recently, SAMHSA cutoff values have been proposed for the screening and confirmation of most illicit drugs in oral fluid.

7.2.3   Analytical Procedures

Oral fluid can be extracted and analyzed as other biological fluids such as blood. In general, there will be less interference from endogenous compounds than with blood or urine. An oral fluid sample collected with a special device usually provides the analyst with a clean specimen. However, a sample collected by spitting contains cell debris, food particles, and strings of mucus; centrifugation is difficult because of the high viscosity. The specimen has to be stored at –4°C and measured as soon as possible because of possible bacterial and fungal growth. The cocaine concentration in saliva, stored in a plastic recipient without addition of citric acid or other stabilizers, remains unaltered at –4°C for 1 week. Freezing of the sample lowers the viscosity substantially so that centrifugation can be performed after thawing. This ensures better handling of the sample and a high stability for most analytes for a long period of time except for THC. Sometimes, the addition of sodium fluoride is reported, e.g., for cannabis and benzodiazepines.

A selection of various chromatographic procedures for drugs of abuse in saliva has been published recently.[2] Quantitation is usually performed with the common GC/MS procedures for drugs of abuse in blood, using electron impact mode and the appropriate deuterated standards. Since oral fluid contains the parent drugs, it is important to add internal standards for the quantification of THC, cocaine, and 6-MAM. Due to the smaller sample volume of an oral fluid specimen in comparison to a blood sample, analytical procedures using MS in chemical ionization mode, tandem MS/MS confirmation, or LC-MS are being developed.

7.2.3.1 Amphetamines

It has been shown recently[1] that the concentrations of amphetamine and MDMA in oral fluid, either obtained by spitting or with a Salivette, exceed the corresponding plasma concentrations 10 to 100 times. Oral contamination after sniffing of the drug can account for only some of the high concentrations during the first hours. Oral fluid samples can also be extracted using the common solid-phase extraction procedures for amphetamines in blood. Proposed SAMHSA screening cutoff values are 50 ng/ml for D-methamphetamine as the target analyte and confirmation cutoff values are 50 ng/ml for D-amphetamine and 50 ng/ml for D-methamphetamine.

A first innovative approach to the problem of low sample volumes is the use of LC-MS-MS with a minimum sample workup. Detection limits are in the range of 2 ng/ml for amphetamine and the designer amphetamines, using only 50 μl of oral fluid.[7] A simple precipitation of the proteins with methanol followed by centrifugation or even dilution of the saliva sample followed by direct injection allows the analyst to process a high number of oral fluid samples with a small sample size.

7.2.3.2 Cannabis

Although there is very little transport for THC from blood to saliva, the usual administration routes (smoking, ingestion) provide detectable levels of THC for several hours with different techniques (HPLC, GC-ECD, GC-MS; limit of quantification 1 to 20 ng/ml). It is obvious that the higher THC concentrations after smoking are due to contamination of the buccal cavity. Calculation of S/P ratios in these circumstances is of little value but the detection of cannabinoids in saliva is a better indication of recent use than detection of the metabolite in urine.

The analytes of interest are THC, cannabinol, and cannabidiol.[1] SAMHSA proposes a screening cutoff of 4 ng/ml for THC as the target analyte for the initial screen and 4 ng/ml of THC in the confirmation analysis. Extraction from an oral fluid sample can be done with a mixture of hexane/ethylacetate (9/1) (v/v) after acidification; derivatization is performed by methylation with tetramethylammoniumhydroxide and iodomethane and subsequent extraction into isooctane. The limit of detection can be increased by pulsed splitless injection of 4 μl of the derivatized extract; linearity is observed in the range 0.5 to 50 ng/ml, with a LOD of 0.1 ng/ml.

It has been demonstrated[8] that the cotton from the Salivette adsorbs 90% of the THC content of a spiked saliva sample. After centrifugation, no THC could be detected in the recovered fluid. Extraction of the dry cotton with methanol or with hexane/ethylacetate (9/1) (v/v) after addition of THC-d$_3$ to the roll resulted in THC concentrations exceeding 100 ng/salivette in drivers showing impairment and providing a positive blood sample.[5]

7.2.3.3 Cocaine

Numerous reports have documented the excretion of cocaine and metabolites in saliva. Cocaine appeared in saliva rapidly following intravenous injection, inhalation, and intranasal administration to volunteers. Contamination of the oral cavity after smoking and sniffing was variable but significant during the first 2 h after dosing. Anhydroecgonine methyl ester was detectable in saliva after smoking, but it was quickly cleared. Benzoylecgonine and ecgonine methyl ester levels were only comparable with cocaine concentrations at times when those had declined to below 100 ng/ml. However, the concentrations of the metabolites will be higher in chronic users. Proposed SAMHSA screening cutoff values are 20 ng/ml for benzoylecgonine as the target analyte, 8 ng/ml for the confirmation.

Generally, cocaine and its metabolites can be detected in saliva using a simple solid-phase extraction procedure: the sample is diluted with acetate buffer pH 4.0 and brought on a conditioned Bond Elut Certify® column; washing is performed with water, buffer pH 4.0, and acetonitrile, and elution occurs with dichloromethane/isopropanol/ammonia (80/20/2) (v/v/v). Replacing the buffer with 0.1 N HCl allows a washing step with methanol instead of acetonitrile because of the ionary binding of benzoylecgonine to the column. Derivatization is often performed with BSTFA but pentafluoropropionic anhydride in combination with pentafluoropropanol provides a better sensitivity for benzoylecgonine. Limits of detection are in the range 1 to 5 ng/ml.

7.2.3.4 Opiates

Heroin is rapidly metabolized to 6-MAM in blood, which is in turn converted to morphine. After intravenous injection, the saliva/plasma ratio for heroin and 6-MAM can be lower than one, depending on the salivary pH. After smoking or sniffing of heroin, the concentrations of the analytes in saliva remain higher than in plasma for 2 to 3 h due to a contamination of the buccal cavity. SAMHSA proposed cutoff values are 40 ng/ml morphine for the initial screen and 40 ng/ml of morphine and 4 ng/ml of 6-MAM for the confirmation. Detection of morphine, codeine, and 6-MAM in saliva is based on similar analytical procedures as for cocaine. However, a washing step with a strong acid will result in the hydrolysis of 6-MAM, so acetate buffer pH 4.0 is preferred for opioids. GC-MS data after silylation show similar results as for the PFP derivatives.

Codeine was detected in oral fluid for 12 to 24 h after oral intake of 30 mg of liquid codeine phosphate depending on the individual and on the collection protocol.[4] After solid-phase extraction, derivatization occurred with TFAA and GC-MS analysis was performed using positive-ion chemical ionization. The limit of quantification was 5 ng/ml, and limit of detection 1.0 ng/ml. Substantially different pharmacokinetic parameters were detected after spitting, with or without stimulation, using a Salivette or the Finger collector containing Accu-Sorb®. Moreover, *in vitro* studies showed that more than 90% of codeine and morphine could be recovered from the Salivette after centrifugation and only 60% from the Finger collector after milking the foam.

REFERENCES

1. Samyn, N. and van Haeren, C., On-site testing of saliva and sweat with Drugwipe, and determination of concentrations of drugs of abuse in saliva, plasma and urine of suspected users, *Int. J. Legal Med.*, 113, 150, 2000.
2. Samyn, N. et al., Analysis of drugs of abuse in saliva, *Forensic Sci. Rev.*, 11, 1, 1999.
3. Kato, K. et al., Cocaine and metabolite excretion in saliva under stimulated and nonstimulated conditions, *J. Anal. Toxicol.*, 17, 338, 1993.
4. O'Neal, C.L. et al., The effects of collection methods on oral fluid codeine concentrations, *J. Anal. Toxicol.*, 24, 536, 2000.
5. Verstraete, A. and Puddu, M., Deliverable D4: evaluation of different roadside drug tests. ROSITA Contract DG VII RO 98-SC.3032, 2000. http://www.rosita.org.
6. Cone, E.J. et al., Oral fluid testing for drugs of abuse: positive prevalence rates by Intercept immunoassay screening and GC-MS/MS confirmation and suggested cut-off concentrations, *J. Anal. Toxicol.*, 26, 541, 2002.
7. Wood, M. et al., Development of a rapid and sensitive method for the quantitation of amphetamines in human plasma and oral fluid by LC-MS-MS, *J. Anal. Toxicol.*, 27, 78, 2003.
8. Kintz, P., Cirimele, V., and Ludes, B., Detection of cannabis in oral fluid (saliva) and forehead wipes (sweat) from impaired drivers, *J. Anal. Toxicol.*, 24, 557, 2000.

7.3 SWEAT

Researchers have known for more than a century[1] that drugs are excreted in sweat. Although it is still poorly understood how nonvolatile chemicals exit the body through the skin, significant advances in sweat analyses have been made during the past few years with the development of the sweat-patch technology.[2] Over a period of 1 to 7 days, sweat will saturate the pad located in the center of the patch and will slowly concentrate, and drugs present in the perspiration fluid will be retained. Sweat appears to offer the advantage of being a non-invasive means of obtaining a cumulative estimate of drug exposure over 1 week. Sweat testing has found applications in monitoring of individuals in drug rehabilitation or in probation/parole programs.[3,4]

The term *sweat-testing* is misleading, because drugs are not only excreted by sweat glands; the excretion of drugs through the skin is also possible by sebaceous glands or transdermal liquid transport.

Sweat is a liquid secreted by the sweat glands, originating deep within the skin. Water (99%) is the major component. Na^+ and Cl^- are the major ions, present in variable concentrations, ranging from 5 to 80 mmol/L. Amino acids, biogenic amines, and vitamins are only present in trace amounts. The pH (4 to 6) is strongly associated with the amount of lactic acid excreted. Apocrine glands secrete a more viscous, cloudy, yellow-white liquid, which is primarily sterile and odorless, and rich in cholesterol (75%), triglycerides, and fatty acids (20%).

Sebaceous glands are associated with hair follicles and located everywhere except on palms and soles; they are particularly abundant on the scalp and forehead. They produce a viscous, yellow-white liquid oily sebum that consists of triglycerides (60%) and wax esters. Excreted sweat and sebum cannot be examined separately.

Sweat secretion is an important homeostatic mechanism for maintaining a constant core body temperature. At temperatures above 31°C, body heat is dissipated by the release of sweat on the skin surface resulting in evaporative heat loss. Between 300 and 700 ml/day of *insensible* sweat is produced over the whole body, likely caused by diffusion through the skin, whereas 2 to 4 L/h of *sensible* sweat may be produced for short periods by extensive exercise. The amount of sweat that is produced is thus affected by body location, ambient temperature, body temperature, and relative humidity of the environment and by emotional, mental, and physical stress. The variability of these factors and the uneven distribution of the sweat glands make it difficult to obtain specimens of sweat systematically.

7.3.1 Sweat Collection

Almost all studies obtain mixed secretions of sweat and sebum, which is incorrectly referred to as sweat. Methods to collect drugs in sweat have included the use of gauze, cotton, towel, or filter paper to absorb sweat and the collection of liquid sweat in rubber gloves or plastic body bags.[2,5] Thermal[6] or pharmacological stimulations such as with pilocarpine[7] were proposed to secrete an unusually large amount of sweat.

Patches, similar to bandages, have been developed to wear for extended periods of time. Early patches were made of absorbent cotton pads sandwiched between a waterproof, polyurethane outer layer and a porous inner layer that is placed against the skin. They have been successfully applied to the detection of ethanol in sweat.[8]

Significant advances have been made during the past few years to develop a non-occlusive sweat collection device, which has been marketed as the PharmChek™ sweat patch. The device consists of an adhesive layer on a thin transparent film of surgical dressing to which a rectangular absorbent pad is attached. The sweat patch acts as a specimen container for nonvolatile and liquid components of sweat, including drugs. Nonvolatile substances from the environment cannot penetrate the transparent film, which is a semipermeable membrane over the pad that allows oxygen, water, and carbon dioxide to pass through the patch, leaving the skin underneath healthy. Over a period of 1 to 10 days, sweat saturates the pad and the water content of the perspiration is volatilized by the body's heat; drugs present in sweat are retained. The region of the skin where the patch will be applied is thoroughly decontaminated with 50% of isopropanol. Subjects can wear one patch with minimal discomfort for at least 1 week. Normal hygiene practices can be achieved. Attempts to remove the patch prematurely are readily visible to personnel trained to monitor the sweat patch.

An alternative collection for epidemiological surveys and for testing of potentially impaired drivers consists of wiping the skin with a cosmetic pad moistened with isopropanol.[9] This procedure allows rapid collection of drugs that may arise both from sweat evaporating on the surface of the skin and from external contamination.

7.3.2 Drug Detection without a Patch Collection

By using some of the early "home-made" collection methods, it was possible to identify various drugs, including methadone, phenobarbital, morphine, cocaine, THC, and methamphetamine.[10] Sweat samples that are collected on gauze or cotton by wiping the surface of the skin are eluted with water and extracted by liquid–liquid methods.

Kidwell et al.[11] investigated an alternative collection of sweat to detect cocaine prevalence in a university population. Sampling of hands and forehead was performed with a cotton pad (skin wipe) moistened with 500 μl of 90 or 70% isopropanol. LODs of the GC-MS-CI confirmatory method were 1.2 to 2.2 ng/wipe for cocaine and benzoylecgonine. A skin swab test may provide a rapid way to gather data on drug use/exposure in a certain population without the embarrassment often felt when obtaining urine samples and the concern for cosmetic damage often expressed when individuals are asked to take hair samples. Skin wipes detect drugs excreted in sweat/sebum more rapidly than a device (patch) that needs to collect pure sweat over an extended period of time.

Sweat, as a forehead specimen, was obtained from injured drivers[9] by wiping the forehead with a commercial pad spiked with 0.5 ml of water/isopropanol (50:50). After extraction of the pad with hexane/ethyl acetate (90:10) and methylation, THC was detectable by GC/MS between 4 and 152 ng/pad, with the risk of environmental contamination. Neither 11-hydroxy-THC nor THC-COOH was identified.

7.3.3 Drug Detection with a Patch Collection

Controlled experimental studies using the Pharmchek™ sweat patch have been performed for cocaine,[2,12] heroin,[13] and amphetamine derivatives, such as methamphetamine,[14] MBDB,[15] and MDMA.[16]

Administration of low doses of cocaine (about 1 to 5 mg) produced detectable amounts of cocaine in sweat.[12] Sweat patches were applied to the back and abdomen of subjects prior to and periodically after drug administration. Before affixing the patches, the skin area was cleaned with an isopropyl alcohol (70%) swab. The absorbent pad was extracted with 2.5 ml of a mixture of 0.1% Triton X-100 in 0.2 M acetate buffer, in the presence of deuterated internal standards. After agitation and centrifugation, the filtered extract solution was purified by SPE and submitted to GC/MS. LODs for cocaine and metabolites were approximately 1 ng/patch. Within- and between-run coefficients of variation were less than or equal to 10%. It was observed that the parent drug was the predominant analyte with cocaine levels (ng/patch) being much higher than the concentrations of the metabolites benzoylecgonine and ecgonine methyl ester. After a single 25-mg IV injection, 42-mg smoked, and 32-mg intranasal administration, cocaine appeared in sweat within 2 h after drug administration and peaked within 24 h with maximum concentrations varying from as low as 15 to as high as 70 ng cocaine/patch. Inter-subject variability was high whereas intra-subject variability was relatively low.

More recent work has focused on the applications of the "Fast Patches" requiring only 30 min for sweat collection because they employ heat-induced sweat stimulation and a larger cellulose pad for increased drug collection.[2] Sweat was collected periodically for 48 h following subcutaneous administration of 75 or 150 mg of cocaine hydrochloride/70 kg. Torso Fast Patches were applied to the abdomen or flank, and Hand-held Fast Patches were affixed to the palm of the nondominant hand and elution of the patch was performed with 0.5 M sodium acetate buffer (pH 4.0). Peak cocaine concentrations (4.5 to 24 h after dosing) ranged from 33 to 3579 ng/patch for the Hand-held Patch and from 22 to 1463 ng/patch for the Torso Fast Patch. Cocaine could be detected for at least 48 h after dosing. Drug concentrations were considerably higher than those reported for the Pharmchek sweat patch. Cocaine concentrations in the Hand-held Patch were more than twofold greater than those found in the Torso Fast Patch.

In a study conducted during a heroin maintenance program, 14 subjects were injected with two or three doses of heroin hydrochloride ranging from 80 to 1000 mg/day. The sweat patches were applied before the first dosage and removed after 24 h, minutes before the next dosage.[13] Patches were extracted with acetonitrile before GC-MS analysis. Concentrations (ng/patch) ranged from 2.1 to 96.3 for heroin, 0 to 24.6 for 6-MAM, and 0 to 11.2 for morphine. There was no correlation between the doses of heroin administered and the concentrations of heroin measured in sweat.

An enzyme immunoassay screening test was described for methamphetamine in eluates of sweat patches, with a cross-reactivity of 144% for MDMA, 30% for amphetamine, and 21% for MDA. Diagnostic sensitivity and specificity were 84.5 and 93.2%, respectively, when comparing to the GC-MS results and applying a cutoff value of 10 ng/ml amphetamine equivalents (using methamphetamine calibrators). The clinical sensitivity and specificity of the overall analysis system were 85 and 100%, respectively, using known methamphetamine dosing of volunteers as the reference standard. Drug doses given were probably lower than a street drug dose.[14]

N-Methyl-1-(3,4-methylenedioxyphenyl)-2-butanamine (MBDB) was detected in the methanolic extract from the Pharmchek sweat patch with concentrations increasing during the first 36 h following a single dose of 100 mg MBDB.[15] Peak concentrations of MBDB and its metabolite 3,4-methylene-dioxyphenyl-2-butanamine (BDB) were 44 and 23 ng/patch, respectively.

In a recent study,[16] sweat was collected for up to 24 h with the PharmChek sweat patches from recreational users of MDMA who received a single 100-mg dose of the drug. After SPE and derivatization, analytes were tested by GC/MS. MDMA was detected in sweat as early as 1.5 h after administration and peaked at 24 h. In all the nine subjects, an on-site test with the Drugwipe immunochemical strip test was positive at 1.5 h.

As a result of the poor passive diffusion from plasma, it was necessary to use very sensitive techniques to detect some drugs in sweat. Ion trap GC/MS/MS was reported as useful for cannabis.[17] THC was identified in sweat of nine cannabis users after wearing a patch for 5 days (4 to 38 ng/patch).[3]

REFERENCES

1. Tachau, H., Uber den Ubergang von Arzneimitteln in der Schweiss, *Arch. Exp. Pathol. Pharmakol.*, 66, 224, 1911.
2. Huestis, M. et al., Sweat testing for cocaine, codeine and metabolites by gas chromatography-mass spectrometry, *J. Chromatogr. B*, 733, 247, 1999.
3. Kintz, P. et al., Sweat testing in opioid users with a sweat patch, *J. Anal. Toxicol.*, 20, 393, 1996.
4. Huestis, M.A. et al., Monitoring opiate use in substance abuse treatment patients with sweat and urine drug testing, *J. Anal. Toxicol.*, 24, 509, 2000.
5. Inoue, T. and Seta, S., Analysis of drugs in unconventional samples, *Forensic Sci. Rev.*, 4, 90, 1992.
6. Fox, R.H. et al., The nature of the increase in sweating capacity produced by heat acclimatization, *J. Physiol.*, 171, 368, 1964.
7. Balabanova, S. et al., Die Bedeutung der Drogenbestimmung in Pilocarpinschweiss für den Nachweis eines zurückliegendes Drogenkonsum, *Beitr. Gerichtl. Med.*, 50, 111, 1992.

8. Phillips, M., Sweat-patch testing detects inaccurate self-reports of alcohol consumption, *Alcohol. Clin. Exp. Res.,* 8, 51, 1984.

9. Kintz, P., Cirimele, V., and Ludes, B., Detection of cannabis in oral fluid (saliva) and forehead wipes (sweat) from impaired drivers, *J. Anal. Toxicol.,* 24, 557, 2000.

10. Kidwell, D.A., Holland, J.C., and Athanaselis, S., Testing for drugs of abuse in saliva and sweat, *J. Chromatogr. B,* 713, 111, 1998.

11. Kidwell, D.A., Blanco, M.A., and Smith, F.P., Cocaine detection in a university population by hair analysis and skin swab testing, *Forensic Sci. Int.,* 84, 75, 1997.

12. Cone, E.J. et al., Sweat testing for heroin, cocaine and metabolites, *J. Anal. Toxicol.,* 18, 298, 1994.

13. Kintz, P. et al., Sweat testing for heroin and metabolites in a heroin maintenance program, *Clin. Chem.,* 43, 736, 1997.

14. Fay, J. et al., Detection of methamphetamine in sweat by EIA and GC-MS, *J. Anal. Toxicol.,* 20, 398, 1996.

15. Kintz, P., Excretion of MBDB and BDB in urine, saliva, and sweat following single oral administration, *J. Anal. Toxicol.,* 21, 570, 1997.

16. Pichini, S. et al., Usefulness of sweat testing for the detection of MDMA after a single-dose administration, *J. Anal. Toxicol.,* 27, 294, 2003.

17. Ehorn, C., Fretthold, D., and Maharaj, M., Ion trap GC/MS/MS for analysis of THC from sweat. In *Proceedings of the TIAFT-SOFT Congress 1994,* Tampa, FL, abstr. 11.

7.4 GENERAL CONCLUSION

It appears that the value of alternative specimen analysis for the identification of drug users is steadily gaining recognition. This can be seen from its growing use in pre-employment screening, in forensic sciences, and in clinical applications.

Saliva and sweat will probably be used in the near future in roadside testing both for epidemiological and screening purposes.

Hair analysis may be a useful adjunct to conventional drug testing in toxicology. Methods for evading urinalysis do not affect hair analysis.

Specimens of saliva or sweat can be more easily obtained with less embarrassment than for urine, and hair can provide a more accurate history of drug use. However, due to a lack of suitable screening tools, costs are too high for routine use (because the analyses require hyphenated chromatographic procedures) but the generated data are extremely helpful to document positive urine cases. These new matrices may find useful applications in the near future, for example, in occupational medicine, roadside testing, or doping control or for law enforcement agencies to document illicit drug use.

CHAPTER 8

Interpreting Alternative Matrix Test Results

Edward J. Cone, Ph.D.,[1] **Angela Sampson-Cone, Ph.D.,**[1] **and Marilyn A. Huestis, Ph.D.**[2*]

[1] ConeChem Research, Severna Park, Maryland

[2] Chemistry and Drug Metabolism Section, Intramural Research Program, National Institute on Drug Abuse, National Institutes of Health, Baltimore, Maryland

CONTENTS

8.1 DRUG TESTING WITH ALTERNATIVE MATRICES

Significant advances in analytical technology over the last decade have provided the analytical toxicologist with the means of testing for drugs of abuse and their metabolites in biologic matrices with ultrahigh sensitivity and specificity. Conclusive identification and measurement of drug resi-

---

* Dr. Huestis contributed to this book in her personal capacity. The views expressed are her own and do not necessarily represent the views of the National Institutes of Health or the U.S. government.

dues in tissues provide important information on an individual's past drug use and exposure history. Such evidence can be important in many types of forensic investigations, e.g., workplace testing, accident investigations, health assessments, diagnosis, treatment and monitoring of drug abusers, and whether drug use was a contributory factor in civil and criminal acts. Many centrally acting drugs produce significant physiologic and behavioral impairment. Frequently, when accidents and loss of life and property occur, medical personnel and toxicologists are called upon to interpret toxicology results. This evidence may support or refute the presence of drug-induced behavioral changes in an individual. In addition, results may indicate whether drug use or drug exposure occurred, the frequency and route of drug use, and the probable level of exposure.

Many different types of biologic specimens can be tested for drug residues. The most frequently collected types are urine, oral fluid, sweat, hair, and blood. Circumstances often determine the type of specimen collected. Whereas post-mortem testing traditionally involves collection of blood and tissue specimens, accident and criminal investigations most frequently involve urine and blood. Workplace and treatment programs generally test urine specimens but, more recently, some have begun to test oral fluid, hair, and sweat as alternative or complementary specimens. Urine and sweat are utilized most commonly in the legal community, e.g., parolee monitoring.

The growing use of different biological specimens presents an array of challenges to those involved in forensic and clinical interpretation. The unique properties of each biological specimen offer somewhat different information on past drug use, and it is becoming increasingly important to understand these differences. In some cases, test results from diverse biological specimens may not agree. The reasons for agreement or disagreement may be complex. Forensic interpretation of specimen test results must be based on an understanding of the chemical and pharmacologic properties of the drug, performance characteristics of analytical methods, and the unique physiological properties of the biological specimens.

This chapter examines the similarities and differences of drug disposition in blood, urine, oral fluid, sweat, and hair specimens. The primary focus is on the chemical and physiologic properties of different specimens and the time course of drug disposition into and out of the biologic matrix. Differences in the physicochemical properties of drugs add to the complexity of interpretation of results. Cocaine is used as the primary model compound in this chapter to illustrate similarities and differences in specimen types. Understanding the "uniqueness" of drug disposition in biological matrices will guide interpretation of similar and disparate results from multiple biological specimens.

8.2 DRUG DISPOSITION IN ALTERNATIVE MATRICES

8.2.1 Circumstances of Drug Exposure

Drug exposure takes place under many different circumstances. Active drug use can involve a variety of routes of administration including oral ingestion, inhalation, parenteral administration, and absorption through skin and mucosal membranes. Passive drug exposure can occur through inhalation of drug smoke, vapor, or dust, by external contamination of the skin, hair, and mucous membranes, and by oral ingestion, e.g., doping. "Active" drug administration is defined generally as intentional drug self-administration by the individual undergoing testing, whereas "passive drug exposure" is defined as unintentional drug exposure. Figure 8.1 provides a graphic illustration of the complex rate processes involved in drug use and exposure and in transfer into and out of different bodily tissues.

Regardless of how drug use or exposure occurs, the consequences of a positive test may be highly detrimental to the donor. Understanding how drugs penetrate and reside in different tissues is an important component in the development and use of biological specimens for drug testing.

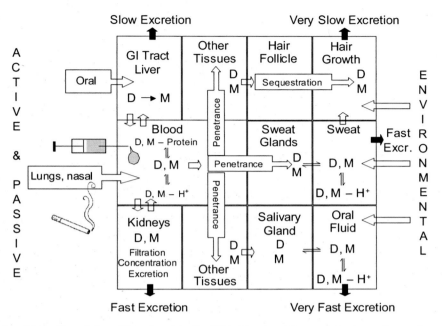

Figure 8.1 Disposition of drug (D) and metabolite (M) in blood and tissues following active and passive drug administration and environmental exposure. The rate of excretion of drug and metabolite is indicated as "Very Fast" (minutes), "Fast" (minutes to hours), "Slow" (hours to days), and "Very Slow (days to months). Abbreviations: D = drug; M = metabolite; D,M–Protein = protein-bound drug or metabolite; D, M–H⁺ = ionized drug or metabolite.

8.2.2 Absorption and Distribution

During the process of penetrance into the body, drugs may be altered by metabolic and chemical processes. Drug metabolism that occurs during absorption, distribution, and elimination processes can result in the production of metabolites with substantially different properties than the parent drug. The pharmacologic activity of metabolites may be enhanced, reduced, or totally eliminated by molecular alteration. Once drug penetrance into the bloodstream occurs, the route of administration becomes less important to drug disposition. Depending on the chemical properties of the drug, metabolism may occur immediately in blood, but more commonly occurs following uptake into the liver and other organs. Arterial blood carries drugs and metabolites to tissues and venous blood transports them away. The transfer of drug and metabolites into tissues is governed by factors such as the chemical properties of the drug, degree of tissue perfusion, protein binding, and characteristics of the tissue membrane. Drug uptake into tissues is generally rapid and equilibrium between highly perfused tissues and venous blood is quickly established. The rate of drug uptake is dependent on the rate of blood flow and the tissue/blood partition coefficient. Accumulation of drug by organ tissues occurs in the extracellular fluid. Some drugs also cross cell membranes into intracellular water. Long-term accumulation of lipid-soluble drugs may occur in fatty tissues.

8.2.3 Clearance and Elimination

Most drugs are cleared from the body through the processes of metabolism and excretion. The liver is the major site of drug metabolism and the type of metabolite produced is based on the enzymatic processes involved. A drug may be subject to hydrolytic cleavage, oxidation, reduction, and conjugation reactions, depending upon its molecular structure. Frequently, these reactions produce metabolites with greater water solubility and less pharmacologic activity than the parent molecule. Such reactions primarily are irreversible and serve to clear active drug from the body. The enhanced water-solubility of metabolites facilitates their removal by excretory processes.

Nonvolatile drugs and metabolites are generally excreted in urine and feces. Other pathways of drug elimination include excretion in bile, saliva, sweat, milk (via lactation), sebum, other bodily fluids, skin (via desquamation), and hair. Some processes, such as excretion into bile and saliva, may be circulatory in nature and reabsorption into the bloodstream may occur during transit through the gastrointestinal tract.

8.3 PHYSIOLOGIC CONSIDERATIONS

8.3.1 Blood

Although blood is infrequently collected in the workplace testing, it is important to understand drug disposition in this matrix. The primary mode of entry of drugs and metabolites into other biological specimens is from blood. On average, an adult human male weighing 70 kg has a blood volume of approximately 5 L. Blood is composed of two parts: plasma, the fluid portion, and formed elements that include blood cells and platelets that are suspended in the fluid. There are more than 60 proteins in plasma that fight infection, transport lipids, and prevent uncontrolled bleeding. Albumin is a major component of plasma protein and undergoes reversible binding with many drugs. Binding to protein has a major influence on a drug's disposition in the body. Protein-bound drug is a large molecular complex that cannot easily cross membranes. An equilibrium exists between protein-bound drug and free drug. Consequently, drug distribution to other tissues is limited to the portion of drug that exists as free unbound drug.

Drug molecules dissolved in blood are transported through a network of fine capillaries to the tissues. The rate of passage of drugs across cell membranes is dependent on the drug's chemical properties and the properties of the cell membrane. Capillary wall membranes are generally highly porous, whereas other membranes may be considerably more complex and less porous, e.g., unique endothelial cells of the blood–brain barrier and the multilayered epithelial cell layers of the dermis and epidermis. Lipid-soluble drugs generally diffuse more easily across membranes than polar, water-soluble drugs, metabolites, and highly protein-bound drugs.

Following drug absorption into the bloodstream, changes in blood concentrations initially reflect changes that occur as a result of distribution and uptake by tissues. At some point after administration, drug concentration in plasma approaches equilibrium with drug sequestered in tissues. Once equilibrium is established, drug concentration declines as a result of metabolic and elimination processes. Most drug concentrations decline at a rate proportional to the amount of drug remaining in the body. The elimination half-life ($t_{1/2}$) of a drug is the time required for the plasma concentration to be reduced by one half. The half-life of a particular drug generally is constant for an individual across a broad range of dosages; however, some drugs exhibit nonlinear kinetics when dosages overwhelm enzymatic and/or elimination processes.

An illustration of the decline of plasma cocaine concentrations of one subject (Subject I) following intravenous, intranasal, and smoked administration of 25 mg of cocaine hydrochloride is shown in Figure 8.2. Cocaine exhibited a plasma half-life of approximately 1.4 h after either intravenous or smoked administration. The decline in cocaine concentration following intranasal administration was biphasic, likely as a result of irregular absorption.

The appearance of drug metabolites in blood is influenced by the route of administration. The liver is the major site of drug metabolism. When drug is ingested, absorption of virtually all drugs is faster from the small intestine than from the stomach due to its large surface area. Therefore, gastric emptying is a rate-limiting step. The intraluminal pH is 4 to 5 in the duodenum, but becomes progressively more alkaline, approaching a pH of 8 in the lower ileum. Absorption of drugs from

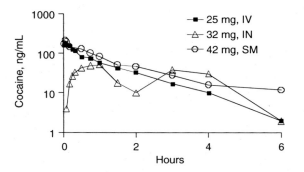

Figure 8.2 Plasma concentrations of cocaine in a subject (Subject I) after intravenous (IV), intranasal (IN), and smoked (SM) administration of single doses of cocaine.

the gastrointestinal tract depends on the drug's ability to pass across intestinal cell membranes and resist destruction in the liver (first-pass effect). In most cases, drugs pass through intestinal membranes by simple diffusion, from an area of high concentration (inside the lumen) to an area of lower concentration (bloodstream). During absorption, significant metabolism may occur, so that only a fraction of the drug may survive intact. Metabolites formed in the liver may be excreted into the feces via bile, or may be released into the bloodstream at the same time as parent drug. Hence, following oral administration, the initial appearance of a metabolite in the bloodstream may be rapid with concentrations sometimes exceeding that of the parent drug.

Drug administration by other routes, e.g., parenteral, transmucosal, and inhalation, may partially or totally bypass the liver and avoid the "first-pass effect" of oral ingestion. Once drug enters the bloodstream, it is immediately distributed to all parts of the body. Approximately 25% of cardiac output is directed to the liver, where drug may be extracted from blood. A variety of processes may affect the ability of the liver to extract drug from blood, including the degree of drug-protein binding (only free, unbound drug may be extracted), solubility of the drug in the hepatic membrane, and metabolic processes occurring in the hepatocytes. Once formed, metabolite(s) may be transported into bile and released into the gastrointestinal tract or be returned to the bloodstream.

In contrast to oral administration, initial appearance of metabolites following parenteral and mucosal administration may be delayed. As an example, Figure 8.3 illustrates the appearance in plasma of cocaine and benzoylecgonine (BZE), a metabolite of cocaine, following intravenous administration. BZE was initially detected at 10 min and rose to a maximum concentration in approximately 1 h following cocaine administration. Thereafter, BZE declined slowly with an approximate half-life of 6.3 h. The rapid formation and slow disappearance of BZE relative to cocaine indicates that elimination, rather than metabolism, was the rate-limiting step for BZE.

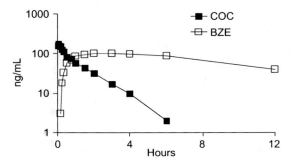

Figure 8.3 Plasma concentrations of cocaine (COC) and benzoylecgonine (BZE) in a subject (Subject I) after intravenous administration of a single 25-mg dose of cocaine.

When the half-life of a metabolite is longer than that of the parent drug, accumulation of the metabolite can occur with repeated, frequent administrations, as occurs with BZE.

8.3.2   Urine

Approximately 20 to 25% of the blood from cardiac output, or 1200 mL/min, goes to the kidney. Blood enters the glomerulus for nonselective filtering of the plasma. Most drug molecules with low molecular weights can pass through the filter forming an ultrafiltrate with a concentration similar to plasma; however, only the unbound portion of the drug can be filtered. Consequently, drugs that are highly bound to plasma protein generally undergo a longer renal elimination period than drugs that are less highly bound. In addition, some drugs undergo active renal secretion, a carrier-mediated system. If the carrier has greater affinity for the free drug than plasma protein, then active renal secretion can increase the rate of elimination despite protein binding.

The glomerulus receives blood first and filters about 125 mL/min of plasma-containing substances with molecular weights less than 20,000 Da. The plasma ultrafiltrate flows through the proximal convoluted tubule, loop of Henle, and distal tubule, and continues through the collecting duct to the bladder. During transit through the nephron, electrolytes, nutrients, and water are reabsorbed and returned to the bloodstream. Excess water and waste products, such as urea, organic substances (including drugs and metabolites), and inorganic substances, are eliminated from the body. Normally, water is reabsorbed reducing urine flow to less than 1 to 2 mL/min. The daily amount and composition of urine vary widely depending upon many factors such as fluid intake, diet, health, drug effects, and environmental conditions. The volume of urine produced by a healthy adult in a 24-h period ranges from 1 to 2 L with a pH range of 5 to 7.5, but normal values outside these limits may be encountered.

As a consequence of extensive water reabsorption in the kidney, drugs and metabolites concentrate in urine. Indeed, if a drug or metabolite exists largely in the unbound state in plasma, its relative concentration could be up to 100 times greater in urine. Figure 8.4 illustrates the concentrations of cocaine and BZE in urine compared to plasma in a subject following intravenous administration of cocaine. Urine cocaine and BZE concentrations were approximately 15 and 100 times higher than in plasma, respectively. These findings suggest that BZE is less highly bound to protein than cocaine or alternately that cocaine is reabsorbed by the kidney during renal excretion. It is important to note that despite the large concentration differences observed in plasma as compared to urine, estimation of elimination half-lives were similar for both drugs in plasma and urine.

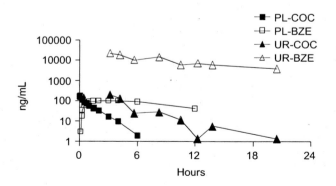

Figure 8.4   Plasma (PL) and urine (UR) concentrations of cocaine (COC) and benzoylecgonine (BZE) in a subject (Subject I) after intravenous administration of a single 25-mg dose of cocaine.

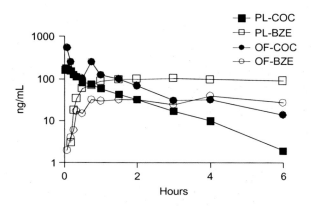

Figure 8.5 Plasma (PL) and oral fluid (OF) concentrations of cocaine (COC) and benzoylecgonine (BZE) in a subject (Subject I) after intravenous administration of a single 25-mg dose of cocaine.

8.3.3 Oral Fluid

Oral fluid is a composite specimen that consists primarily of saliva, gingival crevicular fluid (fluid from the gingival crevice), and cellular debris. The bulk of fluid in the oral cavity originates from the submandibular and parotid glands and gingival crevicular fluid. During resting conditions, about 71% of oral fluid is supplied by the submandibular glands, approximately 25% originates from the parotid glands, and the remaining 4% is produced by other minor glands.[1] With stimulation, the fluid contribution from the parotid gland rises to about 41%. Salivary glands are innervated by both sympathetic and parasympathetic nerves. Saliva production occurs in response to neurotransmitter release. The volume of saliva varies considerably from approximately 500 to 1500 mL/day. The pH of saliva is generally acidic, but may range from 6.0 to 7.8, depending on the rate of saliva flow. As saliva flow increases, levels of bicarbonate increase, thus increasing pH.[2]

Saliva glands, as they are highly perfused, have a high blood flow. Drug in plasma is distributed rapidly to salivary glands and may appear in saliva within minutes of parenteral drug administration.[3] Figure 8.5 illustrates the rapid appearance of cocaine and BZE in oral fluid compared to plasma. Cocaine concentrations in oral fluid generally exceeded those of plasma by a factor of approximately two. The acidic nature of oral fluid compared to plasma and cocaine's lipophilicity favor excretion in oral fluids and account for its higher concentration. BZE also appeared in oral fluid simultaneously with its appearance in plasma, but was present at approximately one third of its plasma concentration. The greater water solubility and lower lipid solubility of BZE apparently accounts for its lower concentration in oral fluid relative to plasma. However, the longer half-life of BZE compared to cocaine leads to prolonged appearance in oral fluid, in this case, exceeding the concentration of cocaine within 4 to 6 h. With multiple cocaine doses and accumulation of BZE, testing for BZE in oral fluid is likely to be of greater utility for the detection of cocaine use.

Other drugs of abuse such as opiates, amphetamines, phencyclidine, and designer drugs that contain basic nitrogen moieties exhibit similar distribution patterns in oral fluid compared to plasma.[4] In contrast, tetrahydrocannabinol (THC), the major active ingredient of cannabis, is a non-nitrogen-containing compound that is highly fat soluble. Deposition of THC in the oral cavity during smoking and oral ingestion most likely accounts for its presence in oral fluid. Fortunately, the residence time of THC in the oral cavity is sufficient for detection over a similar time course as its presence in plasma. Indeed, detection rates for cannabis use by oral fluid testing appear to be similar to those seen in urine testing.[5]

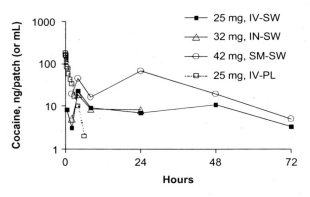

Figure 8.6   Sweat (SW) patch concentrations of cocaine in a subject (Subject I) after intravenous (IV), intranasal (IN), and smoked (SM) administration of single doses of cocaine. For comparison, plasma (PL) cocaine concentrations after IV administration are also shown (dotted line). Sweat patches were applied and removed between each collection point.

8.3.4   Sweat

Sweat secretion is an important mechanism for maintaining a constant core body temperature. Following sympathetic nerve stimulation, sweat is excreted onto the surface of the skin and evaporated to release body heat. Sweat is secreted from two types of sweat glands: eccrine and apocrine glands. These glands originate deep within the skin dermis and terminate in excretory ducts emptying onto the skin or developing hair follicles. Eccrine glands are located on most skin surfaces, while apocrine glands are restricted to skin of the axilla, genitalia, and anus. Water is the primary constituent of sweat accounting for approximately 99% of its bulk,[6] and sodium chloride is the most concentrated solute. Sweat also contains albumin, gamma globulins, waste products, trace elements, drugs, and many other substances found in blood. The rate of sweating is highly dependent on the environmental temperature. Above 31°C, humans begin to sweat and may excrete as much as 3 L/h over short periods of time. The average pH of sweat of resting individuals is reported to be 5.82.[7] Following exercise, the pH increases with increasing flow rates and is reported to be between 6.1 and 6.7.[8] Approximately 50% of sweat is generated by the trunk of the body, 25% from the legs, and the remaining 25% from the head and upper extremities.[8]

Blood flow delivers drug and metabolite to sweat glands, i.e., eccrine and apocrine glands, where diffusion of unbound drug and metabolite occurs through membranes. The rate of passage of drug from blood to sweat via the sweat gland epithelium appears to be proportional to the oil/water partition coefficient of the unionized drug.[9] This process would be facilitated for drugs that demonstrate low degrees of ionization at physiological pH. For lipid-soluble, basic drugs with high pKa values, excretion in sweat is further enhanced by the acidic nature of sweat. For example, cocaine generally is excreted in sweat in higher concentrations than found in plasma. Figure 8.6 illustrates the excretion of cocaine in sweat following intravenous, intranasal, and smoked routes of administration. Plasma cocaine concentrations are also shown following intravenous administration. After 2 h, sweat cocaine concentrations were considerably higher in sweat than in plasma. Despite the longer half-life of BZE, it was present in only trace concentrations (not shown).

8.3.5   Hair

Hair consists of five morphological components: cuticle, cortex, medulla, melanin granules, and cell membrane complex. Each is distinct in morphology and chemical composition. The number of human head hair follicles ranges from 80,000 to 100,000, but this decreases with age.[10] Hair follicles are embedded in the dermis of skin and are highly vascularized to nourish the growing hair root or bulb.[11] The bulb at the base of the follicle contains matrix cells that give rise to the layers of the hair shaft including the cuticle, cortex, and medulla. Matrix cells undergo morphological and structural changes as they move upward during growth to form different layers of the hair shaft. These layers can often be distinguished by qualitative and quantitative differences in their proteins and pigments.[10] Hair is composed of approximately 65 to 95% protein, 1 to 9% lipid, and small quantities of trace elements, polysaccharides, and water.[11] The durability and strength of the hair shaft is determined by the proteins synthesized within the matrix cells. Matrix cells also may acquire pigment or melanin during differentiation into individual layers of hair. The pigment present in hair cells determines the color of the hair shaft. The primary structure of hair consists of two or three α-keratin chains wound into strands called microfibrils. Microfibrils are organized into larger bundles of macrofibrils that comprise the bulk of the cortex. Hair strands are stabilized and shaped by disulfide and hydrogen bonds giving the microfibrils a semi-crystalline structure. Cytochrome P450 and other enzymes have been identified in the hair follicle providing evidence that drug metabolism may occur within this structure; however, as hair becomes fully keratinized in 3 to 5 days, this capability is presumed to be greatly reduced.[11]

A protective layer of epithelial cells called the cuticle surrounds the cortex. The cuticle is the outermost layer of hair, the innermost region is the medulla, and the hair cortex lies between these components. The overlapping cuticle cells protect the cortex from the environment. As hair ages, there is a gradual degeneration of the cuticle along the shaft due to exposure to ultraviolet radiation, chemicals, and mechanical stresses. The cuticle may be partially or totally missing in cases of damaged hair.[12] Hair damaged by cosmetic treatments and/or ultraviolet radiation may influence the deposition and stability of drug in the hair.[13]

Hair follicles continue to grow for a number of years and undergo different phases during a normal growth cycle of several years. Approximately 85 to 90% of the hair is in the anagen or growth phase at any single time.[10] A small portion of mature hair then enters the catagen phase, in which there is a rapid reduction in growth rate. This phase lasts for a period of 2 to 3 weeks and is immediately followed by the telogen phase, the resting phase of hair. During this phase, no growth occurs. Approximately 10 to 15% of head hair is usually in the telogen phase. The hair strand may not be shed for several months prior to replacement by a new strand. Hair growth rates vary according to body location, sex, and age.

Head hair grows at an average rate of 1.3 cm/month although there is some variation according to sex, age, and ethnicity.[14] Mangin and Kintz[15] determined morphine and codeine concentrations in human head, axillary, and pubic hair from heroin overdose cases. Morphine concentrations were highest in pubic hair followed by head and axillary hair. Codeine to morphine ratios ranged from 0.054 to 0.273. The slower growth rate of pubic hair and possible drug contribution from sweat or urine were presumed to account for the differences in drug concentration in the different hair samples.

There are multiple possible pathways for drug incorporation into hair, including (1) passive diffusion from blood into the hair follicle; (2) excretion onto the surface of hair from sweat and sebum; (3) passage from skin to hair; and (4) from external contamination. Henderson[16] suggested that drugs may enter hair from multiple sites, via multiple mechanisms, and at various times during the hair growth cycle. Drugs and their metabolites are distributed throughout the body primarily by passive diffusion from blood. Distribution across membranes is generally facilitated by high lipid solubility, low protein binding, and physicochemical factors that favor the unionized form of the drug in blood. Diffusion of drug from arterial blood capillaries to matrix cells in the base of the follicle is considered a primary means for drug deposition in hair. Presumably, drug binds to components in the matrix and to pigments. As the cells elongate and age, they gradually die and coalesce, forming the nonliving hair fiber. Drug that may be present is embedded in the hair matrix.

Drug entering hair via blood from the capillary plexus of the follicle is not detectable by standard hair cutting methods until hair grows to the skin surface. With daily sampling of beard, the time course for morphine and codeine to appear was reported to be in the range of 7 to 10 days.[17] In controlled dosing studies with cocaine and codeine, these drugs were detectable in "unwashed"

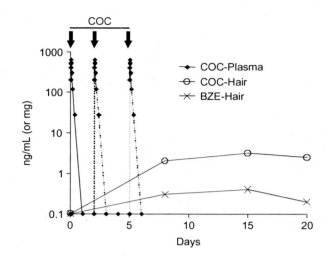

Figure 8.7   Cocaine (COC) concentrations in plasma and cocaine and benzoylecgonine (BZE) concentrations in human head hair after 150 mg/kg cocaine administration by the subcutaneous route. Data represent the mean average for five subjects. Three separate cocaine administrations occurred on days 0, 2, and 5. The dotted lines for cocaine plasma represent simulated plasma concentrations (not measured). (Data adapted from References 18 and 19.)

human head hair approximately 8 days after the first drug administration.[18] Peak drug concentrations occurred in approximately 15 days. BZE concentrations were detectable over the same time course at approximately one tenth the concentration of cocaine. Figure 8.7 illustrates mean concentrations of cocaine in plasma and cocaine and BZE in hair.[18,19] Of special interest in these studies was the finding that solvent washes removed 50 to 55% of cocaine and codeine in the first hair specimens collected after drug dosing, but minimal drug was removed by solvent washing of specimens collected in later weeks. It was concluded that early deposition of drug on hair by sweat accounted for the differences observed in solvent washing and that multiple mechanisms may be involved in drug deposition in hair. Environmental contamination also has been reported to a potential route of cocaine entry into hair.[20,21]

8.4 PHARMACOKINETIC, CHEMICAL, AND ANALYTICAL CONSIDERATIONS

The pattern of disposition of drugs in alternative matrices is influenced by time of collection relative to dosing, frequency of drug use, chemical and physiological differences between alternative matrices, and chemical differences between drugs and between parent drug and metabolites. The relative abundance of parent drug and metabolites is a dynamic process, especially over the first 12 to 24 h after drug administration. Generally, parent drug concentrations are highest in biological tissues during this time. Some drugs, such as heroin, may be metabolized so quickly that their presence in tissues like blood and urine may be extremely short-lived. In contrast, heroin sequestration in hair and in sweat (sweat patch) may have a stabilizing effect and preserve the molecule intact for detection for long intervals after use. Metabolic processes and distribution to tissues proceed rapidly, thereby altering relative abundances of parent drug and metabolites. Multiple dosing also can exert significant influences on the pattern of drug disposition. Frequently, drug metabolites have a longer half-life than the parent, resulting in accumulation of metabolites to concentrations exceeding those of the parent drug. In some cases, metabolites are the only detectable analytes. Table 8.1 presents a summary of metabolic and chemical influences on drug and metabolite distribution in alternative matrices.

Table 8.1 Metabolic and Chemical Influences on Drug and Metabolite Distribution in Alternate Matrices

Drug	Blood	Urine	Oral Fluid	Sweat	Hair
Cannabis	THC highly lipid soluble and rapidly distributed to tissues resulting in a short detection period; THCCOOH formed rapidly from THC; significant accumulation of THCCOOH may occur from frequent use	THCCOOH conjugated and rapidly excreted; accumulation after multiple dosing significantly extends detection period	THC sequestered in oral cavity after smoking and ingestion providing basis for detection; sequestered THC in oral mucosa may influence blood THC	THC found in low concentrations (<2 ng/patch) after controlled drug administration of up to 27 mg; THC in sweat from frequent users can be >100 ng/patch or Drugwipe	THC and THCCOOH sequestered in hair matrix; identification of THCCOOH considered evidence of active use; THCCOOH present in extremely low concentrations
Cocaine	COC rapidly converted to BZE and EME; hydrolysis to BZE continues to occur after collection unless chemically stabilized	COC and BZE rapidly excreted; rate of COC excretion may be influenced by urine pH; BZE has longer half-life	Acidic pH of oral fluid concentrates COC compared to blood; acidic pH enhances stability of COC; BZE concentration < blood	Acidic pH of sweat enhances stability of COC, but hydrolysis to BZE in collection device may occur	Cocaine binding to hair matrix enhances stability; hydrolysis to BZE may occur; BZE/COC ratio >0.1; cocaethylene or norcocaine considered evidence of active use
Heroin/morphine	HER rapidly hydrolyzed to 6-AM and MOR; lipid solubility of heroin and 6-AM facilitates rapid uptake into tissues as compared to morphine; MOR extensively conjugated	MOR and conjugated MOR rapidly excreted; 6-AM excreted in small amounts for a few hours after heroin administration	Acidic pH of oral fluid enhances 6-AM concentration compared to blood; pH enhances stability of 6-AM	Acidic pH of sweat enhances stability of heroin and 6-AM, but hydrolysis to MOR in collection device may occur	HER and 6-AM binding to hair matrix enhances stability; hydrolysis to MOR may occur
Codeine	COD rapidly and extensively metabolized by conjugation; forms small amounts of active metabolite (MOR) considered responsible for effects; genetic variability in O-demethylation activity	COD, conjugated COD, MOR, and conjugated MOR rapidly excreted	Acidic pH of oral fluid enhances COD concentration compared to blood	Acidic pH of sweat likely enhances COD concentration	COD bound to hair matrix; no further metabolic changes considered likely to occur
PCP	PCP highly lipid soluble and rapidly distributed to tissues	Rate of excretion influenced by urine pH	Acidic pH of oral fluid likely enhances concentration compared to blood	Acidic pH of sweat likely enhances concentration	PCP bound to hair matrix; no further metabolic changes considered likely to occur
Amphetamine	AMP lipid soluble and rapidly distributed to tissues	Rate of excretion influenced by urine pH	Acidic pH of oral fluid enhances concentration	Acidic pH of sweat likely enhances concentration	AMP bound to hair matrix; no further metabolic changes considered likely to occur
Methamphetamine	METH lipid soluble and rapidly distributed to tissues; limited metabolism to AMP	Rate of excretion influenced by urine pH; presence of AMP generally required for confirmation	Acidic pH of oral fluid enhances concentration	Acidic pH of sweat enhances concentration	METH bound to hair matrix; no further metabolic changes considered likely to occur

Abbreviations: THC = tetrahydrocannabinol; THCCOOH = 11-nor-9-carboxy-Δ9-tetrahydrocannabinol; COC = cocaine; BZE = benzoylecgonine; EME = ecgonine methyl ester; HER = heroin; 6-AM = 6-acetylmorphine; MOR = morphine; COD = codeine; PCP = phencyclidine; METH = methamphetamine; AMP = amphetamine.

Development of drug testing methodologies for alternative matrices involves consideration of the spectrum of target analyte(s) and acceptable cutoff concentration(s) for useful detection windows. Table 8.2 provides a summary of target analytes, pharmacokinetic considerations, references for analysis of each drug, and for interpretation of results. Each alternative matrix has unique characteristics that influence test outcomes. Some generalities are useful as guides to alternative matrices testing:

Detection methods for drugs in urine:
  • Frequently target end-stage metabolites
  • Frequently require a hydrolysis step to measure conjugated metabolites
  • Demand only moderate to high sensitivity due to the concentrating ability of the kidneys

Detection methods for drugs in oral fluid and blood:
  • Have similarities in target analyte distribution and time course of detection, except for cannabis where detection in oral fluid is primarily due to sequestration of THC in the oral cavity
  • Frequently measure active parent drug
  • Require ultrasensitive methodology due to low concentrations

Detection methods for drugs in sweat and hair:
  • Generally have similar distribution of parent drug and metabolites
  • Require ultrasensitive methodology due to low concentrations

8.5 INTERPRETATION OF ALTERNATIVE MATRICES

8.5.1 Time Course of Detection in Alternative Matrices

The time course of detection of drugs and metabolites in alternative matrices varies from hours to months depending on the type of biological specimen being tested. Generally, drug detection times follow the following order: blood < oral fluid < urine < sweat < hair. A general scheme for drug detection in different matrices is shown in Figure 8.8. Drug detection in blood and oral fluid generally follows the same time course. Drugs of abuse are generally detected in blood and oral fluid from 1 to 3 days depending on the pattern of drug usage and cutoff concentration.

Single doses of most drugs of abuse are detected in urine for 1 to 3 days.[22] Multiple dosing extends the detection window for cocaine, opiates, and amphetamines by 1 to 2 days, whereas multiple dosing of cannabis and phencyclidine may result in positive urine test results for up to 27 days.[23] A short lag time (hours) may be observed for drug appearance in urine following drug administration. For example, in some individuals, delays up to 4 h have been noted for the appearance of cannabinoids in urine following marijuana smoking.[24]

Studies on sweat testing with the Sweat Patch showed that it takes approximately 2 to 8 h following drug administration for sufficient drug to be deposited in the patch for detection. Additional cocaine use, while the patch is being worn, results in further deposition. Cocaine use 24 to 48 h prior to application of the patch may also produce positive results.[25] However, drug use just prior to patch removal is not likely to be detected because of the delay in appearance of drug in sweat.

Multiple mechanisms for drug entry into hair may account for the confusion regarding the beginning of the drug detection window.[18,26,27] Drug excreted in sweat may appear on hair within hours after use, but also may be more difficult to detect if washing procedures efficiently remove the bulk of drug residue. A period of hair growth estimated to be 3 to 5 days[14] must occur for drug in hair to appear at the skin surface, thereby accounting for the "lag" period (days) in its detection window.

Changes in analytical sensitivity and specificity, e.g., antibody changes in the screening assay, may result in enhanced or diminished detection times for all types of specimens. Also, administrative changes in recommended cutoff concentrations could have a similar effect.

Table 8.2 Relative Abundances of Target Analytes in Alternative Matrices, Pharmacokinetic Considerations, and Selected References to Aid Interpretation of Results

Drug	Blood/Plasma/Serum	Urine	Oral Fluid	Sweat	Hair
Cannabis					
Target analytes	THCCOOH > THC	THCCOOH-G > THCCOOH	THC	THC	THC > THCCOOH
PK effects	Collection time, multiple dosing and long $t_{1/2}$ of THCCOOH alters THCCOOH/THC ratio[44–50]	Multiple dosing and long $t_{1/2}$ of THCCOOH produces accumulation and prolongs detection[23,51–68]	Multiple dosing may produce accumulation and prolonged detection[35,69–73]	Multiple dosing may produce accumulation and prolonged detection[71,74–78]	Multiple dosing may be necessary for detection; metabolite identification important to eliminate environmental contamination[79–86]
Cocaine					
Target analytes	BZE > COC > EME > NCOC; CE (with ethanol)	BZE > COC	BZE > COC	COC > BZE	COC > BZE > NCOC; CE (with ethanol)
PK effects	Collection time, multiple dosing and long $t_{1/2}$ of BZE alters BZE/COC ratio; COC in combination with ethanol forms CE[3,19,87–104]	Collection time, multiple dosing and long $t_{1/2}$ of BZE alters COC/BZE ratio; COC in combination with ethanol forms CE[25,60,87,89–91,93–95,105–117]	Collection time, multiple dosing and long $t_{1/2}$ of BZE alters COC/BZE ratio; COC in combination with ethanol forms CE[92,94–96,105,118–124]	Collection time, multiple dosing and long $t_{1/2}$ of BZE alters COC/BZE ratio; COC in combination with ethanol forms CE[25,32,76,125,126]	Metabolite identification, e.g., NCOC, important to eliminate environmental contamination; presence of CE indicates systemic COC and ethanol exposure[18,83,87,97,118,127–140]
Heroin/morphine					
Target analytes	MOR-G > MOR > 6-AM > HER	MOR-G > MOR > 6-AM > HER	6-AM MOR > HER	6-AM MOR > HER	6-AM > MOR
PK effects	Collection time, multiple dosing and long $t_{1/2}$ of MOR alters MOR-G/MOR/6-AM ratios; HER and 6-AM frequently not detected[17,96,97,141–148]	Collection time, multiple dosing and long $t_{1/2}$ of MOR alters MOR-G/MOR/6-AM ratios; HER and 6-AM frequently not detected[17,87,143,144,149–163]	6-AM and MOR most frequently detected; HER may also be detected[17,75,87,96,141,142,150,159,164–166]	6-AM and MOR most frequently detected; HER also may be detected[74–76,125,126,163,167–169]	Multiple dosing may be necessary for detection; 6-AM and MOR most frequently detected, but HER also may be detected[17,83,87,97,127,128,130–135,139,142,143,165,170–174]
Codeine					
Target analytes	COD-G > COD > MOR > NCOD	COD-G > COD > MOR > NCOD	COD > MOR	COD > MOR	COD > MOR> NCOD
PK effects	Collection time, multiple dosing may alter COD-G/COD/MOR/NCOD ratios[17,19,142,175–178]	Collection time, multiple dosing may alter COD-G/COD/MOR/NCOD ratios; MOR may exceed COD late in excretion phase[17,149,160,179–184]	Collection time, multiple dosing may alter COD/MOR ratio; MOR may not be detected[17,123,142,166,175,177,178,185–188]	Collection time, multiple dosing may alter COD/MOR ratio[32,123,189,190]	Multiple dosing may be necessary for detection[17,18,127,142,191–197]

Continued

Table 8.2　Relative Abundances of Target Analytes in Alternative Matrices, Pharmacokinetic Considerations, and Selected References to Aid Interpretation of Results (Continued)

Drug	Blood/Plasma/Serum	Urine	Oral Fluid	Sweat	Hair
PCP					
Target analytes	PCP	PCP > HO-PCP	PCP	PCP	PCP
PK effects	Multiple dosing may substantially increase detection time[198-207]	Multiple dosing may substantially increase detection time[202,208-211]	Multiple dosing may substantially increase detection time[212,213]	Multiple dosing may substantially increase detection time[76]	Multiple dosing may be necessary for detection[136,214-219]
Amphetamine					
Target analytes	AMP	AMP	AMP	AMP	AMP
PK effects	Multiple dosing may increase detection time[204,220-224]	Multiple dosing may increase detection time[221,225-242]	Multiple dosing may increase detection time[35,36,75,223,224,241,243,244]	Multiple dosing may increase detection time[35,75,243,245]	Multiple dosing may be necessary for detection[36,38,82,97,128,131,222,243,246-255]
Methamphetamine					
Target analytes	METH > AMP	METH > AMP	METH > AMP	METH > AMP	METH > AMP
PK effects	Multiple dosing may increase detection time[204,220,221,243,256-260]	Collection time, multiple dosing may alter METH/AMP ratio[221,226,229,231-233,235,237,241-243,258,261-268]	Collection time, multiple dosing may alter METH/AMP ratio[241,243,244,269]	Collection time, multiple dosing may alter METH/AMP ratio[243,269,270]	Multiple dosing may be necessary for detection; AMP present in lower amounts[243,246,249,251-255,269,271-275]
MDA/MDMA/MDEA					
Target analytes					
PK effects	MDMA demonstrates nonlinear rise in plasma levels with increasing dose; poor CYP2D6 metabolizers may be at risk of toxicity[276-281]	Rapidly excreted primarily as parent drug, N-desmethyl-metabolites, and hydroxyl-metabolites[233,279-285]	Acidic pH of oral fluid enhances concentration compared to blood[277,281]	Detectable in sweat for 2–12 h after administration[286,287]	MDA/MDMA/MDEAMDMA concentrations usually exceed MDA (metabolite)[135,219,247-249,251-253,288-292]

Abbreviations: THC = tetrahydrocannabinol; THCCOOH = 11-nor-9-carboxy-Δ9-tetrahydrocannibinol; THCCOOH-G = 11-nor-9-carboxy-Δ9-tetrahydrocannibinol-glucuronine; COC = cocaine; BZE = benzoylecgonine; EME = ecgonine methyl ester; NCOC = norcocaine; CE = cocaethylene; MOR-G = morphine glucuronide; MOR = morphine; HER = heroin; 6-AM = 6-acetylmorphine; COD-G = codeine glucuronide; COD = codeine; PCP = phencyclidine; HO-PCP = hydroxyphencyclidine; METH = methamphetamine; AMP = amphetamine; MDA = methylenedioxyamphetamine; MDMA = methylenedioxymethamphetamine; MDEA = methylenedioxyethylamphetamine.

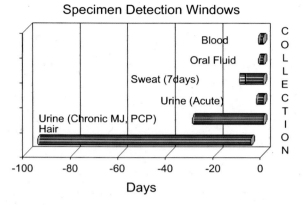

Figure 8.8   General detection times for drugs in blood, oral fluid, urine, sweat, and hair. Lighter shaded area of sweat indicates possible detection of drug use that occurred 24–48 h prior to application of sweat patch.

8.5.2 Multiple Specimen Testing

Continued development, approval, and use of alternative matrices in drug testing programs may present interesting problems in interpretation of results. In the past, it was rare to have test results from more than one type of specimen, particularly in workplace testing. The financial costs of testing more than one type of biological specimen almost certainly will force most drug testing programs to choose a single type of specimen best suited to their needs; however, there are likely to be instances when individuals and even entire drug testing programs decide to test more than one biological specimen. Already, individuals informed of positive test results may request additional testing, e.g., an individual who tested positive in a urine testing program may choose to undergo hair testing. Medical Review Officers may also request additional testing if there is reasonable doubt concerning the validity of a test result. Post-accident testing calls for the highest scrutiny regarding the potential role of drugs as causative factors; hence, multiple specimens may become the norm in this type of testing arena. In addition, drug testing authorities may want information on test comparability, e.g., urine vs. oral fluid testing, prior to switching to a new matrix. Consequently, it is quite likely that there will be many instances where multiple test results from different biological matrices will be collected and require interpretation.

8.5.3 Guidance in Interpretation of Alternative Matrices Results

Considerable guidance information is available for interpretation of positive urine test results.[22,28] Although there is less information available for alternative matrices, a number of reviews are helpful in this regard.[1,2,4,29-43] When multiple test results from different matrices are available, interpretation may be considerably more complex, particularly for disparate results. There are a number of legitimate reasons why one might obtain disparate results. Each type of biological specimen has unique physiological and chemical properties that may alter the pattern of drug disposition. For example, renal excretion processes favor elimination of water-soluble metabolites, whereas excretion of drug into oral fluid favors parent drugs capable of rapid passive diffusion across membranes. Sweat excretion processes also appear to favor parent drug. The acidic nature of oral fluid and sweat favors excretion and trapping of drugs containing basic nitrogen moieties. Although the complex mechanisms for drug binding to hair pigment and proteins have not been fully elucidated, it is clear that these binding processes exhibit greater affinity for drugs containing basic nitrogen moieties. Residence times in each matrix also differ substantially, yielding wide variability in windows of detection.

The many differences in physiological and chemical properties of alternative matrices undoubtedly result in production of occasional disparate test results when more than one type of biological specimen is utilized for testing. There are many possible explanations for disparate results. For example, testing two different matrices, e.g., urine and oral fluid, may result in two equivalent or two disparate results. When one considers the possible combinations of results that could arise from testing of blood, urine, oral fluid, sweat, and hair, there are 20 possible disparate results if only two matrices are tested. The different scenarios of disparate results for two specimens and some of the possible explanations are depicted in Table 8.3. It will become the responsibility of the Medical Review Officer and forensic toxicologists to provide interpretation of such results. Clearly, much information must be available to provide a scientific basis for the interpretation of alternative matrices' results.

Table 8.3   Disparate Results from Testing Two Different Biological Matrices and Possible Explanations

Scenario	Blood	Urine	Oral Fluid	Sweat	Hair	Possible Explanations for Disparate Results
			Matrix			
1	Positive	Negative				Time of urine collection too close to time of drug use
2	Positive		Negative			Highly protein-bound drugs may be poorly distributed to oral fluid, e.g., benzodiazepines
3	Positive			Negative		Low drug dose; sampling time outside detection "window"
4	Positive				Negative	Low drug dose; low binding affinity to hair matrix; hair treatments, e.g., bleaching, straighteners; sampling time outside detection "window"
5	Negative	Positive				Long interval after dosing; concentration effect by kidney
6		Positive	Negative			Long interval after dosing; concentration effect by kidney; highly protein-bound drug; sampling time outside detection "window"
7		Positive		Negative		Concentration effect by kidney; sampling time outside detection "window"
8		Positive			Negative	Concentration effect by kidney; low dose; low binding affinity to hair matrix; sampling time outside detection "window"
9	Negative		Positive			Insufficient time for drug absorption; "depot" effect
10		Negative	Positive			Insufficient time for drug absorption, metabolism, and excretion; "depot" effect
11			Positive	Negative		Insufficient time for drug absorption, metabolism and excretion; "depot" effect; sampling time outside detection "window"
12			Positive		Negative	Low drug dose; low binding affinity to hair matrix; insufficient time for drug absorption, metabolism, and excretion; sampling time outside detection "window"
13	Negative			Positive		Sampling time outside detection "window"
14		Negative		Positive		Sampling time outside detection "window"
15			Negative	Positive		Sampling time outside detection "window"
16				Positive	Negative	Low drug dose; low binding affinity to hair matrix; insufficient time for drug absorption, metabolism, and excretion; sampling time outside detection "window"
17	Negative				Positive	Sampling time outside detection "window"
18		Negative			Positive	Sampling time outside detection "window"
19			Negative		Positive	Sampling time outside detection "window"
20				Negative	Positive	Sampling time outside detection "window"

REFERENCES

1. R. Haeckel. Factors influencing the saliva/plasma ratio of drugs. *Ann. N.Y. Acad. Sci.* 694: 128–142 (1993).
2. N. Samyn, A. Verstraete, C. van Haeren, and P. Kintz. Analysis of drugs of abuse in saliva. *Forensic Sci. Rev.* 11: 1–19 (1999).
3. E.J. Cone. Pharmacokinetics and pharmacodynamics of cocaine. *J. Anal. Toxicol.* 19: 459–478 (1995).
4. E.J. Cone. Saliva testing for drugs of abuse. *Ann. N.Y. Acad. Sci.* 694: 91–127 (1993).
5. E.J. Cone, L. Presley, M. Lehrer, W. Seiter, M. Smith, K. Kardos, D. Fritch, S. Salamone, and R.S. Niedbala. Oral fluid testing for drugs of abuse: positive prevalence rates by Intercept™ immunoassay screening and GC–MS–MS confirmation and suggested cutoff concentrations. *J. Anal. Toxicol.* 26: 541–546 (2002).

6. S. Robinson and A.H. Robinson. Chemical composition of sweat. *Psychol. Rev.* 34: 202–220 (1954).

7. W.C. Randall. The physiology of sweating. *Am. J. Phys. Med.* 32: 292–318 (1953).

8. D. Doran, J. Terney, M. Varano, and S. Ware. A study of the pH of perspiration from male and female subjects exercising in the gymnasium. *J. Chem. Educ.* 70: 412–414 (1993).

9. H.L. Johnson and H.I. Maibach. Drug excretion in human eccrine sweat. *J. Invest. Dermatol.* 56: 182–188 (1971).

10. L. Potsch. On physiology and ultrastructure of human hair. In *Hair Analysis in Forensic Toxicology: Proceedings of the 1995 International Conference and Workshop*, R.A. de Zeeuw, I. Al Hosani, S. Al Munthiri, and A. Maqbool, Eds. Organizing Committee of the Conference, Abu Dhabi, 1995, 1–27.

11. M.R. Harkey. Anatomy and physiology of hair. *Forensic Sci. Int.* 63: 9–18 (1993).

12. *Human Hair Volume I: Fundamentals and Methods for Measurement of Elemental Composition,* CRC Press, Boca Raton, FL, 1988, 1–88.

13. L. Potsch, R. Aderjan, G. Skopp, and M. Herbold. Stability of opiates in the hair fibers after exposures to cosmetic treatment and UV radiation. In *Proceedings of the 1994 JOINT TIAFT/SOFT International Meeting*, V. Spiehler, Ed. TIAFT/SOFT Joint Congress, 1994, 65–72.

14. M. Saitoh, M. Uzuka, and M. Sakamoto. Rate of hair growth. In *Advances in Biology of Skin,* Vol. IX. *Hair Growth*, W. Montagna and R.L. Dobson, Eds. Pergamon, Oxford, 1969, 183–201.

15. P. Mangin and P. Kintz. Variability of opiates concentrations in human hair according to their anatomical origin: head, axillary and pubic regions. *Forensic Sci. Int.* 63: 77–83 (1993).

16. G.L. Henderson. Mechanisms of drug incorporation into hair. *Forensic Sci. Int.* 63: 19–29 (1993).

17. E.J. Cone. Testing human hair for drugs of abuse. I. Individual dose and time profiles of morphine and codeine in plasma, saliva, urine, and beard compared to drug-induced effects on pupils and behavior. *J. Anal. Toxicol.* 14: 1–7 (1990).

18. R.E. Joseph, Jr., K.M. Hold, D.G. Wilkins, D.E. Rollins, and E.J. Cone. Drug testing with alternative matrices. II. Mechanisms of cocaine and codeine disposition in hair. *J. Anal. Toxicol.* 23: 396–408 (1999).

19. R.E. Joseph, J.M. Oyler, A.T. Wstadik, C. Ohuoha, and E.J. Cone. Drug testing with alternative matrices. I. Pharmacological effects and disposition of cocaine and codeine in plasma, sebum, and stratum corneum. *J. Anal. Toxicol.* 22: 6–17 (1998).

20. W.L. Wang and E.J. Cone. Testing human hair for drugs of abuse. IV. Environmental cocaine contamination and washing effects. *Forensic. Sci. Int.* 70: 39–51 (1995).

21. G. Romano, N. Barbera, and I. Lombardo. Hair testing for drugs of abuse: evaluation of external cocaine contamination and risk of false positives. *Forensic Sci. Int.* 123: 119–129 (2001).

22. M. Vandevenne, H. Vandenbussche, and A. Verstraete. Detection time of drugs of abuse in urine. *Acta Clin. Belg.* 55: 323–333 (2000).

23. A. Smith-Kielland, B. Skuterud, and J. Morland. Urinary excretion of 11-nor-9-carboxy-delta9-tetrahydrocannabinol and cannabinoids in frequent and infrequent drug users. *J. Anal. Toxicol.* 23: 323–332 (1999).

24. R.S. Niedbala, K.W. Kardos, D.F. Fritch, S. Kardos, T. Fries, J. Waga, J. Robb, and E.J. Cone. Detection of marijuana use by oral fluid and urine analysis following single-dose administration of smoked and oral marijuana. *J. Anal. Toxicol.* 25: 289–303 (2001).

25. K.L. Preston, M.A. Huestis, C.J. Wong, A. Umbricht, B.A. Goldberger, and E.J. Cone. Monitoring cocaine use in substance-abuse-treatment patients by sweat and urine testing. *J. Anal. Toxicol.* 23: 313–322 (1999).

26. D.A. Kidwell and D.L. Blank. Mechanisms of incorporation of drugs into hair and the interpretation of hair analysis data. In *Hair Testing for Drugs of Abuse: International Research on Standards and Technology,* E.J. Cone, M.J. Welch, and M.B. Grigson Babecki, Eds. NIH Pub. 95-3727, National Institute on Drug Abuse, Rockville, MD, 1995, 19–90.

27. M.R. Harkey and G.L. Henderson. Hair analysis for drugs of abuse. In *Advances in Biomedical Analytical Toxicology,* Vol. II, R.C. Baselt, Ed. Biomedical Publishers, Chicago, 1989, 298–329.

28. Medical Review Officer Manual for Federal Workplace Drug Testing Programs, W.F. Vogl and D.M. Bush, Eds. Department of Health and Human Services, Washington, D.C., 1997, 1–72.

29. B. Caddy. Saliva as a specimen for drug analysis. In *Advances in Analytical Toxicology*, R.C. Baselt, Ed. Biomedical Publications, Foster City, CA, 1984, 198–254.

30. Y.H. Caplan and B.A. Goldberger. Alternative specimens for workplace drug testing. *J. Anal. Toxicol.* 25: 396–399 (2001).

31. Y. Gaillard and G. Pepin. Testing hair for pharmaceuticals. *J. Chromatogr. B* 733: 231–246 (1999).
32. M.A. Huestis, J.M. Oyler, E.J. Cone, A.T. Wstadik, D. Schoendorfer, and R.E. Joseph, Jr. Sweat testing for cocaine, codeine and metabolites by gas chromatography-mass spectrometry. *J. Chromatogr. B Biomed. Sci. Appl.* 733: 247–264 (1999).
33. O.R. Idowu and B. Caddy. A review of the use of saliva in the forensic detection of drugs and other chemicals. *J. Forensic Sci. Soc.* 22: 123–135 (1982).
34. D.A. Kidwell, E.H. Lee, and S.F. DeLauder. Evidence for bias in hair testing and procedures to correct bias. *Forensic Sci. Int.* 107: 39–61 (2000).
35. D.A. Kidwell, J.C. Holland, and S. Athanaselis. Testing for drugs of abuse in saliva and sweat. *J Chromatogr. B* 713: 111–135 (1998).
36. P. Kintz and N. Samyn. Use of alternative specimens: drugs of abuse in saliva and doping agents in hair. *Ther. Drug Monit.* 24: 239–246 (2002).
37. P. Mangin. Drug analyses in nonhead hair. In *Drug Testing in Hair*, P. Kintz, Ed. CRC Press, Boca Raton, FL, 1996, 279–287.
38. M.R. Moeller. Hair analysis as evidence in forensic cases. *Ther. Drug Monit.* 18: 444–449 (1996).
39. J.C. Mucklow, M.R. Bending, G.C. Kahn, and C.T. Dollery. Drug concentration in saliva. *Clin. Pharmacol. Ther.* 24: 563–570 (1978).
40. H. Sachs. Forensic applications of hair analysis. In *Drug Testing in Hair*, P. Kintz, Ed. CRC Press, Boca Raton, FL, 1996, 211–222.
41. W. Schramm, R.H. Smith, P.A. Craig, and D.A. Kidwell. Drugs of abuse in saliva: a review. *J. Anal. Toxicol.* 16: 1–9 (1992).
42. G. Skoand and L. Potsch. Perspiration versus saliva — basic aspects concerning their use in roadside drug testing. *Int. J. Leg. Med.* 112: 213–221 (1999).
43. V. Spiehler. Hair analysis by immunological methods from the beginning to 2000. *Forensic Sci. Int.* 107: 249–259 (2000).
44. S.J. Heishman, M.A. Huestis, J.E. Henningfield, and E.J. Cone. Acute and residual effects of marijuana: profiles of plasma THC levels, physiological, subjective, and performance measures. *Pharmacol. Biochem. Behav.* 37: 561–565 (1990).
45. E.J. Cone and M.A. Huestis. Relating blood concentrations of tetrahydrocannabinol and metabolites to pharmacologic effects and time of marijuana usage. *Ther. Drug Monit.* 15: 527–532 (1993).
46. M.A. Huestis, J.E. Henningfield, and E.J. Cone. Blood cannabinoids. I. Absorption of THC and formation of 11-OH-THC and THCCOOH during and after smoking marijuana. *J. Anal. Toxicol.* 16: 276–282 (1992).
47. M.A. Huestis, J.E. Henningfield, and E.J. Cone. Blood cannabinoids. II. Models for the prediction of time of marijuana exposure from plasma concentrations of delta-9-tetrahydrocannabinol (THC) and 11-nor-9-carboxy-delta-9-tetrahydrocannabinol (THCCOOH). *J. Anal. Toxicol.* 16: 283–290 (1992).
48. M.A. Huestis, A.H. Sampson, B.J. Holicky, J.E. Henningfield, and E.J. Cone. Characterization of the absorption phase of marijuana smoking. *Clin. Pharmacol. Ther.* 52: 31–41 (1992).
49. D.E. Moody, L.F. Rittenhouse, and K.M. Monti. Analysis of forensic specimens for cannabinoids. I. Comparison of RIA and GC/MS analysis of blood. *J. Anal. Toxicol.* 16: 297–301 (1992).
50. D.E. Moody, K.M. Monti, and D.J. Crouch. Analysis of forensic specimens for cannabinoids. II. Relationship between blood delta-9-tetrahydrocannabinol and blood and urine 11-nor-delta-9-tetrahydrocannabinol-9-carboxylic acid concentrations. *J. Anal. Toxicol.* 16: 302–306 (1992).
51. M.A. Huestis and E.J. Cone. Urinary excretion half-life of 11-nor-9-carboxy-delta9-tetrahydrocannabinol in humans. *Ther. Drug Monit.* 20: 570–576 (1998).
52. P.M. Kemp, I.K. Abukhalaf, J.E. Manno, B.R. Manno, D.D. Alford, M.E. McWilliams, F.E. Nixon, M.J. Fitzgerald, R.R. Reeves, and M.J. Wood. Cannabinoids in humans. II. The influence of three methods of hydrolysis on the concentration of THC and two metabolites in urine. *J. Anal. Toxicol.* 19: 292–298 (1995).
53. P.C. Painter, J.H. Evans, J.D. Greenwood, and W.W. Fain. Urine cannabinoids monitoring. *Diagn. Clin. Testing* 27: 29–33 (1989).
54. G.M. Ellis, M.A. Mann, B.A. Judson, N.T. Schramm, and A. Tashchian. Excretion patterns of cannabinoid metabolites after last use in a group of chronic users. *Clin. Pharmacol. Ther.* 38: 572–578 (1985).

55. P.M. Kemp, I.K. Abukhalaf, J.E. Manno, B.R. Manno, D.D. Alford, and G.A. Abusada. Cannabinoids in humans. I. Analysis of delta-9-tetrahydrocannabinol and six metabolites in plasma and urine using GC-MS. *J. Anal. Toxicol.* 19: 285–291 (1995).

56. M.A. Huestis, J.M. Mitchell, and E.J. Cone. Lowering the federally mandated cannabinoid immunoassay cutoff increases true-positive results. *Clin. Chem.* 40: 729–733 (1994).

57. E.J. Cone and R.E. Johnson. Contact highs and urinary cannabinoid excretion after passive exposure to marijuana smoke. *Clin. Pharmacol. Ther.* 40: 247–254 (1986).

58. R.J. Bastiani. Urinary cannabinoid excretion patterns. In *The Cannabinoids: Chemical, Pharmacologic, and Therapeutic Aspects*, S. Agurell, W.L. Dewey, and R.E. Willette, Eds. Academic Press, Orlando, FL, 1984, 263–280.

59. E.J. Cone, R.E. Johnson, B.D. Paul, L.D. Mell, and J. Mitchell. Marijuana-laced brownies: behavioral effects, physiologic effects, and urinalysis in humans following ingestion. *J. Anal. Toxicol.* 12: 169–175 (1988).

60. E.J. Cone, R. Lange, and W.D. Darwin. *In vivo* adulteration: excess fluid ingestion causes false-negative marijuana and cocaine urine test results. *J. Anal. Toxicol.* 22: 460–473 (1998).

61. A. Costantino, R.H. Schwartz, and P. Kaplan. Hemp oil ingestion causes positive urine tests for delta9-tetrahydrocanabinol carboxylic acid. *J. Anal. Toxicol.* 21: 482–485 (1997).

62. M.A. Huestis, J.M. Mitchell, and E.J. Cone. Urinary excretion profiles of 11-nor-9-carboxy-Δ9-tetrahydrocannabinol in humans after single smoked doses of marijuana. *J. Anal. Toxicol.* 20: 441–452 (1996).

63. M.A. Huestis, J.M. Mitchell, and E.J. Cone. Detection times of marijuana metabolites in urine by immunoassay and GC-MS. *J. Anal. Toxicol.* 19: 443–449 (1995).

64. M.A. Huestis and E.J. Cone. Differentiating new marijuana use from residual drug excretion in occasional marijuana users. *J. Anal. Toxicol.* 22: 445–454 (1998).

65. D.S. Isenschmid and Y.H. Caplan. Incidence of cannabinoids in medical examiner urine specimens. *J. Forensic Sci.* 33(6): 1421–1431 (1988).

66. R.C. Meatherall and R.J. Warren. High urinary cannabinoids from a hashish body packer. *J. Anal. Toxicol.* 17: 439–440 (1993).

67. C. Tomaszewski, M. Kirk, E. Bingham, B. Saltzman, R. Cook, and K. Kulig. Urine toxicology screens in drivers suspected of driving while impaired from drugs. *Clin. Toxicol.* 34(1): 37–44 (1996).

68. R.A. Gustafson, B. Levine, P.R. Stout, K.L. Klette, M.P. George, E.T. Moolchan, and M.A. Huestis. Urinary cannabinoid detection times after controlled oral administration of delta9-tetrahydrocannabinol to humans. *Clin. Chem.* 49: 1114–1124 (2003).

69. M.A. Huestis, S. Dickerson, and E. J. Cone. Can saliva THC levels be correlated to behavior? In *American Academy of Forensic Science*, Fittje Brothers, Colorado Springs, 1992, 190.

70. N. Fucci, N. De Giovanni, M. Chiarotti, and S. Scarlata. SPME-GC analysis of THC in saliva samples collected with "EPITOPE" device. *Forensic Sci. Int.* 119: 318–321 (2001).

71. P. Kintz, V. Cirimele, and B. Ludes. Detection of cannabis in oral fluid (saliva) and forehead wipes (sweat) from impaired drivers. *J. Anal. Toxicol.* 24: 557–561 (2000).

72. D.B. Menkes, R.C. Howard, G.F.S. Spears, and E.R. Cairns. Salivary THC following cannabis smoking correlates with subjective intoxication and heart rate. *Psychopharmacology* 103: 277–279 (1991).

73. A. Jehanli, S. Brannan, L. Moore, and V.R. Spiehler. Blind trials of an onsite saliva drug test for marijuana and opiates. *J. Forensic Sci.* 46: 1214–1220 (2001).

74. P. Kintz, A. Tracqui, P. Mangin, and Y. Edel. Sweat testing in opioid users with a sweat patch. *J. Anal. Toxicol.* 20: 393–397 (1996).

75. N. Samyn and C. van Haeren. On-site testing of saliva and sweat with Drugwipe and determination of concentrations of drugs of abuse in saliva, plasma and urine of suspected users. *Int. J. Legal Med.* 113: 150–154 (2000).

76. D.J. Crouch, R.F. Cook, J.V. Trudeau, D.C. Dove, J.J. Robinson, H.L. Webster, and A.A. Fatah. The detection of drugs of abuse in liquid perspiration. *J. Anal. Toxicol.* 25: 625–627 (2001).

77. L.K. Thompson and E.J. Cone. Determination of delta-9-tetrahydrocannabinol in human blood and saliva by high-performance liquid chromatography with amperometric detection. *J. Chromatogr.* 421: 91–97 (1987).

78. P. Kintz. Drug testing in addicts: a comparison between urine, sweat, and hair. *Ther. Drug Monit.* 18: 450–455 (1996).

79. V. Cirimele, H. Sachs, P. Kintz, and P. Mangin. Testing human hair for cannabis. III. Rapid screening procedure for the simultaneous identification of delta9-tetrahydrocannabinol, cannabinol, and cannabidiol. *J. Anal. Toxicol.* 20: 13–16 (1996).

80. D. Wilkins, H. Haughey, E. Cone, M.A. Huestis, R. Foltz, and D. Rollins. Quantitative analysis of THC, 11-OH-THC, and THCCOOH in human hair by negative ion chemical ionization mass spectrometry. *J. Anal. Toxicol.* 19: 483–491 (1995).

81. T. Cairns, D.J. Kippenberger, H. Scholtz, and W.A. Baumgartner. Determination of carboxy-THC in hair by mass spectrometry. In *Hair Analysis in Forensic Toxicology: Proceedings of the 1995 International Conference and Workshop*, R.A. de Zeeuw, I. Al Hosani, S. Al Munthiri, and A. Maqbool, Eds. Organizing Committee of the Conference, Abu Dhabi, 1995, 185–193.

82. V. Cirimele. Cannabis and amphetamine determination in human hair. In *Drug Testing in Hair*, P. Kintz, Ed. CRC Press, Boca Raton, FL, 1996, 181–189.

83. C. Jurado, M.P. Gimenez, M. Menendez, and M. Repetto. Simultaneous quantification of opiates, cocaine and cannabinoids in hair. *Forensic Sci. Int.* 70: 165–174 (1995).

84. P. Kintz, V. Cirimele, and P. Mangin. Testing human hair for cannabis. II. Identification of THC-COOH by GC-MS-NCI as a unique proof. *J. Forensic Sci.* 40: 619–622 (1995).

85. P. Kintz, V. Cirimele, and P. Mangin. Testing human hair for cannabis. Identification of natural ingredients of cannabis sativa metabolites of THC. In *Hair Analysis in Forensic Toxicology: Proceedings of the 1995 International Conference and Workshop*, R.A. de Zeeuw, I. Al Hosani, S. Al Munthiri, and A. Maqbool, Eds. Organizing Committee of the Conference, Abu Dhabi, 1995, 194–202.

86. S. Strano-Rossi and M. Chiarotti. Solid-phase microextraction for cannabinoids analysis in hair and its possible application to other drugs. *J. Anal. Toxicol.* 23: 7–10 (1999).

87. W.L. Wang, W.D. Darwin, and E.J. Cone. Simultaneous assay of cocaine, heroin and metabolites in hair, plasma, saliva and urine by gas chromatography-mass spectrometry. *J. Chromatogr.* 660: 279–290 (1994).

88. A.J. Jenkins, R.M. Keeenan, J.E. Henningfield, and E.J. Cone. Correlation between pharmacological effects and plasma cocaine concentrations after smoked administration. *J. Anal. Toxicol.* 26: 382–392 (2002).

89. R.H. Williams, J.A. Maggiore, S.M. Shah, T.B. Erickson, and A. Negrusz. Cocaine and its major metabolites in plasma and urine samples from patients in an urban emergency medicine setting. *J. Anal. Toxicol.* 24: 478–481 (2000).

90. E.T. Shimomura, G.D. Hodge, and B.D. Paul. Examination of postmortem fluids and tissues for the presence of methylecgonidine, ecgonidine, cocaine, and benzoylecgonine using solid-phase extraction and gas chromatography-mass spectrometry. *Clin. Chem.* 47: 1040–1047 (2001).

91. M.W. Linder, G.M. Bosse, M.T. Henderson, G. Midkiff, and R. Valdes. Detection of cocaine metabolite in serum and urine: frequency and correlation with medical diagnosis. *Clin. Chim. Acta* 295: 179–185 (2000).

92. E.T. Moolchan, E.J. Cone, A. Wstadik, M.A. Huestis, and K.L. Preston. Cocaine and metabolite elimination patterns in chronic cocaine users during cessation: plasma and saliva analysis. *J. Anal. Toxicol.* 24: 458–466 (2000).

93. J.C. Garriott, R.G. Rodriguez, and J.L. Castorena. Cocaine metabolites in urine and blood after ingestion of Health Inca Tea. *SWAFS* (1988).

94. R.A. Jufer, A. Wstadik, S.L. Walsh, B.S. Levine, and E.J. Cone. Elimination of cocaine and metabolites in plasma, saliva, and urine following repeated oral administration to human volunteers. *J. Anal. Toxicol.* 24: 467–477 (2000).

95. W. Schramm, P.A. Craig, R.H. Smith, and G.E. Berger. Cocaine and benzoylecgonine in saliva, serum, and urine. *Clin. Chem.* 39: 481–487 (1993).

96. A.J. Jenkins, J.M. Oyler, and E.J. Cone. Comparison of heroin and cocaine concentrations in saliva with concentrations in blood and plasma. *J. Anal. Toxicol.* 19: 359–374 (1995).

97. R. Kronstrand, R. Grundin, and J. Jonsson. Incidence of opiates, amphetamines, and cocaine in hair and blood in fatal cases of heroin overdose. *Forensic Sci. Int.* 92: 29–38 (1998).

98. D.N. Bailey. Serial plasma concentrations of cocaethylene, cocaine, and ethanol in trauma victims. *J. Anal. Toxicol.* 17: 79–83 (1993).

99. G.W. Hime, W.L. Hearn, S. Rose, and J. Cofino. Analysis of cocaine and cocaethylene in blood and tissues by GC-NPD and GC-ion trap mass spectrometry. *J. Anal. Toxicol.* 15: 241–245 (1991).

100. M. Perez-Reyes and A.R. Jeffcoat. Ethanol/cocaine interaction: cocaine and cocaethylene plasma concentrations and their relationship to subjective and cardiovascular effects. *Life Sci.* 51: 553–563 (1992).

101. P.R. Puopolo, P. Chamberlin, and J.G. Flood. Detection and confirmation of cocaine and cocaethylene in serum emergency toxicology specimens. *Clin. Chem.* 38: 1838–1842 (1992).

102. R.M. Smith. Ethyl esters of arylhydroxy- and arylhydroxymethoxycocaines in the urines of simultaneous cocaine and ethanol users. *J. Anal. Toxicol.* 8: 38–42 (1984).

103. J. Sukbuntherng, A. Walters, H.-H. Chow, and M. Mayersohn. Quantitative determination of cocaine, cocaethylene (ethylcocaine), and metabolites in plasma and urine by high-performance liquid chromatography. *J. Pharm. Sci.* 84(7): 799–804 (1995).

104. A.H.B. Wu, T.A. Onigbinde, K.G. Johnson, and G.H. Wimbish. Alcohol-specific cocaine metabolites in serum and urine of hospitalized patients. *J. Anal. Toxicol.* 16: 132–136 (1992).

105. J. Leonard, M. Doot, C. Martin, and W. Airth-Kindree. Correlation of buccal mucosal transudate collected with a buccal swab and urine levels of cocaine. *J. Addict. Dis.* 13: 27–33 (1994).

106. M.A. ElSohly, D.F. Stanford, and H.N. ElSohly. Coca tea and urinalysis for cocaine metabolites. *J. Anal. Toxicol.* 10: 256–256 (1986).

107. A.J. Jenkins, T. Llosa, I. Montoya, and E.J. Cone. Identification and quantitation of alkaloids in coca tea. *Forensic Sci. Int.* 77: 179–189 (1996).

108. K.L. Preston, D.H. Epstein, E.J. Cone, A.T. Wtsadik, M.A. Huestis, and E.T. Moolchan. Urinary elimination of cocaine metabolites in chronic cocaine users during cessation. *J. Anal. Toxicol.* 26: 393–400 (2002).

109. K.L. Preston, K. Silverman, C.R. Schuster, and E.J. Cone. Assessment of cocaine use with quantitative urinalysis and estimation of new uses. *Addiction* 92: 717–727 (1997).

110. K.M. Kampman, A.I. Alterman, J.R. Volpicelli, I. Maany, E.S. Muller, D.D. Luce, E.M. Mulholland, A.F. Jawad, G.A. Parikh, F.D. Mulvaney, R.M. Weinrieb, and C.P. O'Brien. Cocaine withdrawal symptoms and initial urine toxicology results predict treatment attrition in outpatient cocaine dependence treatment. *Psychol. Addict. Behav.* 15: 52–59 (2001).

111. D.M. Jacobson, R. Berg, G.F. Grinstead, and J.R. Kruse. Duration of positive urine for cocaine metabolite after ophthalmic administration: implications for testing patients with suspected Horner syndrome using ophthalmic cocaine. *Am. J. Ophthalmol.* 131: 742–747 (2001).

112. E.J. Cone, A. Tsadik, J. Oyler, and W.D. Darwin. Cocaine metabolism and urinary excretion after different routes of administration. *Ther. Drug Monit.* 20: 556–560 (1998).

113. E.J. Cone, S.L. Menchen, B.D. Paul, L.D. Mell, and J. Mitchell. Validity testing of commercial urine cocaine metabolites assays. I. Assay detection times, individual excretion patterns, and kinetics after cocaine administration to humans. *J. Forensic Sci.* 34: 15–31 (1989).

114. J. Ambre, T.I. Ruo, J. Nelson, and S. Belknap. Urinary excretion of cocaine, benzoylecgonine, and ecgonine methyl ester in humans. *J. Anal. Toxicol.* 12: 301–306 (1988).

115. F.K. Rafla and R.L. Epstein. Identification of cocaine and its metabolites in human urine in the presence of ethyl alcohol. *J. Anal. Toxicol.* 3: 59–63 (1979).

116. R. de La Torre, M. Farre, J. Ortuno, J. Cami, and J. Segura. The relevance of urinary cocaethylene following the simultaneous administration of alcohol and cocaine. *J. Anal. Toxicol.* 15: 223 (1991).

117. D.L. Phillips, I.R. Tebbett, and R.L. Bertholf. Comparison of HPLC and GC-MS for measurement of cocaine and metabolites in human urine. *J. Anal. Toxicol.* 20: 305–308 (1996).

118. F.P. Smith and D.A. Kidwell. Cocaine in hair, saliva, skin swabs, and urine of cocaine users' children. *Forensic Sci. Int.* 83: 179–189 (1996).

119. E.J. Cone, J. Oyler, and W.D. Darwin. Cocaine disposition in saliva following intravenous, intranasal, and smoked administration. *J. Anal. Toxicol.* 21: 465–475 (1997).

120. K. Kato, M. Hillsgrove, L. Weinhold, D.A. Gorelick, W.D. Darwin, and E.J. Cone. Cocaine and metabolite excretion in saliva under stimulated and nonstimulated conditions. *J. Anal. Toxicol.* 17: 338–341 (1993).

121. R.S. Niedbala, K. Kardos, T. Fries, A. Cannon, and A. Davis. Immunoassay for detection of cocaine/metabolites in oral fluids. *J. Anal. Toxicol.* 25: 62–68 (2001).

122. E.J. Barbieri, G.J. DiGregorio, A.P. Ferko, and E.K. Ruch. Rat cocaethylene and benzoylecgonine concentrations in plasma and parotid saliva after the administration of cocaethylene. *J. Anal. Toxicol.* 18: 60–61 (1994).

123. P. Kintz, V. Cirimele, and B. Ludes. Codeine testing in sweat and saliva with the Drugwipe. *Int. J. Leg. Med.* 111: 82–84 (1998).

124. P. Campora, A.M. Bermejo, M.J. Tabernero, and P. Fernandez. Quantitation of cocaine and its major metabolites in human saliva using gas chromatography-positive chemical ionization-mass spectrometry (GC-PCI-MS). *J. Anal. Toxicol.* 27: 270–274 (2003).

125. M.A. Huestis, E.J. Cone, C.J. Wong, K. Silverman, and K.L. Preston. Efficacy of sweat patches to monitor cocaine and opiate use during treatment. Problems of Drug Dependence 1997: Proceedings of the 59th Annual Scientific Meeting, NIDA Monograph 187, p. 223, NIH Pub. No. 98-43051, National Institute of Drug Abuse, Rockville, MD (1997).

126. E.J. Cone, M.J. Hillsgrove, A.J. Jenkins, R.M. Keenan, and W.D. Darwin. Sweat testing for heroin, cocaine, and metabolites. *J. Anal. Toxicol.* 18: 298–305 (1994).

127. K.M. Hold, D.G. Wilkins, D.E. Rollins, R.E. Joseph, and E.J. Cone. Simultaneous quantitation of cocaine, opiates, and their metabolites in human hair by positive ion chemical ionization gas chromatography-mass spectrometry. *J. Chromatogr. Sci.* 36: 125–130 (1998).

128. M.R. Moeller, P. Fey, and R. Wennig. Simultaneous determination of drugs of abuse (opiates, cocaine and amphetamine) in human hair by GC/MS and its application to a methadone treatment program. *Forensic Sci. Int.* 63: 185–206 (1993).

129. E.J. Cone, D. Yousefnejad, W.D. Darwin, and T. Maquire. Testing human hair for drugs of abuse. II. Identification of unique cocaine metabolites in hair of drug abusers and evaluation of decontamination procedures. *J. Anal. Toxicol.* 15: 250–255 (1991).

130. D. Garside and B.A. Goldberger. Determination of cocaine and opioids in hair. In *Drug Testing in Hair*, P. Kintz, Ed. CRC Press, Boca Raton, FL, 1996, 151–177.

131. P. Kintz and P. Mangin. Determination of gestational opiate, nicotine, benzodiazepine, cocaine and amphetamine exposure by hair analysis. *J. Forensic Sci. Soc.* 33: 139–142 (1993).

132. S. Magura, R.C. Freeman, Q. Siddiqi, and D.S. Lipton. The validity of hair analysis for detecting cocaine and heroin use among addicts. *Int. J. Addict.* 27: 51–69 (1992).

133. G. Pepin and Y. Gaillard. Concordance between self-reported drug use and findings in hair about cocaine and heroin. *Forensic Sci. Int.* 84: 37–41 (1997).

134. S. Strano-Rossi, A.B. Barrera, and M. Chiarotti. Segmental hair analysis for cocaine and heroin abuse determination. *Forensic Sci. Int.* 70: 211–216 (1995).

135. F. Tagliaro, R. Valentini, G. Manetto, F. Crivellente, G. Carli, and M. Marigo. Hair analysis by using radioimmunoassay, high-performance liquid chromatography and capillary electrophoresis to investigate chronic exposure to heroin, cocaine and/or ecstasy in applicants for driving licenses. *Forensic Sci. Int.* 107: 121–128 (2000).

136. T. Cairns, D.J. Kippenberger, and A.M. Gordon. Hair analysis for detection of drugs of abuse. In *Handbook of Analytical Therapeutic Drug Monitoring and Toxicology*, S.H.Y. Wong and I. Sunshine, Eds. CRC Press, Boca Raton, FL, 1997, 237–251.

137. A.C. Gruszecki, C.A. Robinson, Jr., J.H. Embry, and G.G. Davis. Correlation of the incidence of cocaine and cocaethylene in hair and postmortem biologic samples. *Am. J. Forensic Med. Pathol.* 21: 166–171 (2000).

138. K. Janzen. Concerning norcocaine, ethylbenzoylecgonine, and the identification of cocaine use in human hair. *J. Anal. Toxicol.* 16: 402 (1992).

139. J. Segura, C. Stramesi, A. Redon, M. Ventura, C.J. Sanchez, G. Gonzalez, L. San, and M. Montagna. Immunological screening of drugs of abuse and gas chromatographic-mass spectrometric confirmation of opiates and cocaine in hair. *J. Chromatogr. B Biomed. Sci. Appl.* 724: 9–21 (1999).

140. G.L. Henderson, M.R. Harkey, C. Zhou, R.T. Jones, and P. Jacob III. Incorporation of isotopically labeled cocaine and metabolites into human hair. 1. Dose–response relationships. *J. Anal. Toxicol.* 20: 1-12 (1996).

141. E.A. Kopecky, S. Jacobson, J. Klein, B. Kapur, and G. Koren. Correlation of morphine sulfate in blood plasma and saliva in pediatric patients. *Ther. Drug Monit.* 19: 530–534 (1997).

142. W. Piekoszewski, E. Janowska, R. Stanaszek, J. Pach, L. Winnik, B. Karakiewicz, and T. Kozielec. Determination of opiates in serum, saliva and hair addicted persons. *Przegl. Lek.* 58: 287–289 (2001).

143. M.R. Moeller and C. Mueller. The detection of 6-monoacetylmorphine in urine, serum and hair by GC/MS and RIA. *Forensic Sci. Int.* 70: 125–133 (1995).

144. L.W. Hayes, W.G. Krasselt, and P.A. Mueggler. Concentrations of morphine and codeine in serum and urine after ingestion of poppy seed. *Clin. Chem.* 33: 806–808 (1987).

145. A.J. Jenkins, R.M. Keenan, J.E. Henningfield, and E.J. Cone. Pharmacokinetics and pharmacodynamics of smoked heroin. *J. Anal. Toxicol.* 18: 317–330 (1994).

146. B.A. Goldberger, W.D. Darwin, T.M. Grant, A.C. Allen, Y.H. Caplan, and E.J. Cone. Measurement of heroin and its metabolites by isotope-dilution electron-impact mass spectrometry. *Clin. Chem.* 39: 670–675 (1993).

147. M.J. Burt, J. Kloss, and F.S. Apple. Postmortem blood free and total morphine concentrations in medical examiner cases. *J. Forensic Sci.* 46: 1138–1142 (2001).

148. G. Ceder and A.W. Jones. Concentration ratios of morphine to codeine in blood of impaired drivers as evidence of heroin use and not medication with codeine. *Clin. Chem.* 47: 1980–1984 (2001).

149. L.H. Yong and N.T. Lik. The human urinary excretion pattern of morphine and codeine following the consumption of morphine, opium, codeine and heroin. *Bull. Narcotics* 29: 45–74 (1977).

150. G.S. Yacoubian, Jr., E.D. Wish, and D.M. Perez. A comparison of saliva testing to urinalysis in an arrestee population. *J. Psychoactive Drugs* 33: 289–294 (2001).

151. S.J. Mule and G.A. Casella. Rendering the "poppy-seed defense" defenseless: identification of 6-monoacetylmorphine in urine by gas chromatography/mass spectroscopy. *Clin. Chem.* 34: 1427–1430 (1988).

152. M.L. Smith, E.T. Shimomura, J. Summers, B.D. Paul, D. Nichols, R. Shippee, A.J. Jenkins, W.D. Darwin, and E.J. Cone. Detection times and analytical performance of commercial urine opiate immunoassays following heroin administration. *J. Anal. Toxicol.* 24: 522–529 (2000).

153. A.C. Spanbauer, S. Casseday, D. Davoudzadeh, K.L. Preston, and M.A. Huestis. Detection of opiate use in a methadone maintenance treatment population with the CEDIA 6-acetylmorphine and CEDIA DAU opiate assays. *J. Anal. Toxicol.* 25: 515–519 (2001).

154. E.J. Cone, R. Jufer, and W.D. Darwin. Forensic drug testing for opiates. VII. Urinary excretion profile of intranasal (snorted) heroin. *J. Anal. Toxicol.* 20: 379–391 (1996).

155. E.J. Cone, P. Welch, J.M. Mitchell, and B.D. Paul. Forensic drug testing for opiates. I. Detection of 6-acetylmorphine in urine as an indicator of recent heroin exposure; drug and assay considerations and detection times. *J. Anal. Toxicol.* 15: 1–7 (1991).

156. H.J.G.M. Derks, K.V. Twillert, and G. Zomer. Determination of 6-acetylmorphine in urine as a specific marker for heroin abuse by high-performance liquid chromatography with fluorescence detection. *Anal. Chim. Acta* 170: 13–20 (1985).

157. C.L. O'Neal and A. Poklis. The detection of acetylcodeine and 6-acetylmorphine in opiate positive urines. *Forensic Sci. Int.* 95: 1–10 (1998).

158. J.M. Mitchell, B.D. Paul, P. Welch, and E.J. Cone. Forensic drug testing for opiates. II. Metabolism and excretion rate of morphine in humans after morphine administration. *J. Anal. Toxicol.* 15: 49–53 (1991).

159. I.M. Speckl, J. Hallbach, W.G. Guder, L.V. Meyer, and T. Zilker. Opiate detection in saliva and urine — a prospective comparison by gas chromatography-mass spectrometry. *J. Toxicol. Clin. Toxicol.* 37: 441–445 (1999).

160. E.J. Cone, S. Dickerson, B.D. Paul, and J.M. Mitchell. Forensic drug testing for opiates. V. Urine testing for heroin, morphine, and codeine with commercial opiate immunoassays. *J. Anal. Toxicol.* 17: 156–164 (1993).

161. J. Fehn and G. Megges. Detection of O6-monoacetylmorphine in urine samples by GC/MS as evidence for heroin use. *J. Anal. Toxicol.* 9: 134–138 (1985).

162. G. Fritschi and W.R. Prescott, Jr. Morphine levels in urine subsequent to poppy seed consumption. *Forensic Sci. Int.* 27: 111–117 (1985).

163. M.A. Huestis, E.J. Cone, C.J. Wong, A. Umbricht, and K.L. Preston. Monitoring opiate use in substance abuse treatment patients with sweat and urine drug testing. *J. Anal. Toxicol.* 24: 509–521 (2000).

164. L. Presley, M. Lehrer, W. Seiter, D. Hahn, B. Rowland, M. Smith, K.W. Kardos, D. Fritch, S. Salamone, R.S. Niedbala, and E.J. Cone. High prevalence of 6-acetylmorphine in morphine-positive oral fluid specimens. *Forensic Sci. Int.* 133: 22–25 (2003).

165. J. Jones, K. Tomlinson, and C. Moore. The simultaneous determination of codeine, morphine, hydrocodone, hydromorphone, 6-acetylmorphine, and oxycodone in hair and oral fluid. *J. Anal. Toxicol.* 26: 171–175 (2002).

166. R.S. Niedbala, K. Kardos, J. Waga, D. Fritch, L. Yeager, S. Doddamane, and E. Schoener. Laboratory analysis of remotely collected oral fluid specimens for opiates by immunoassay. *J. Anal. Toxicol.* 25: 310–315 (2001).

167. P. Kintz, R. Brenneisen, P. Bundeli, and P. Mangin. Sweat testing for heroin and metabolites in a heroin maintenance program. *Clin. Chem.* 43: 736–739 (1997).

168. J.A. Levisky, D.L. Bowerman, W.W. Jenkins, and S.B. Karch. Drug deposition in adipose tissue and skin: evidence for an alternative source of positive sweat patch tests. *Forensic Sci. Int.* 110: 35–46 (2000).

169. D.A. Kidwell and F.P. Smith. Susceptibility of PharmChek drugs of abuse patch to environmental contamination. *Forensic Sci. Int.* 116: 89–106 (2001).

170. S. Darke, W. Hall, S. Kaye, J. Ross, and J. Duflou. Hair morphine concentrations of fatal heroin overdose cases and living heroin users. *Addiction* 97: 977–984 (2002).

171. B. Ahrens, F. Erdmann, G. Rochholz, and H. Schutz. Detection of morphine and monoacetylmorphine (MAM) in human hair. *J. Anal. Chem.* 344: 559–560 (1992).

172. E.J. Cone, W.D. Darwin, and W.L. Wang. The occurrence of cocaine, heroin and metabolites in hair of drug abusers. *Forensic. Sci. Int.* 63: 55–68 (1993).

173. P. Kintz, C. Jamey, V. Cirimele, R. Brenneisen, and B. Ludes. Evaluation of acetylcodeine as a specific marker of illicit heroin in human hair. *J. Anal. Toxicol.* 22: 425–429 (1998).

174. B.A. Goldberger, Y.H. Caplan, T. Maguire, and E.J. Cone. Testing human hair for drugs of abuse. III. Identification of heroin and 6-acetylmorphine as indicators of heroin use. *J. Anal. Toxicol.* 15: 226–231 (1991).

175. C.L. O'Neal, D.J. Crouch, D.E. Rollins, A. Fatah, and M.L. Cheever. Correlation of saliva codeine concentrations with plasma concentrations after oral codeine administration. *J. Anal. Toxicol.* 23: 452–459 (1999).

176. M.K. Brunson and J.F. Nash. Gas-chromatographic measurement of codeine and norcodeine in human plasma. *Clin. Chem.* 21: 1956–1960 (1975).

177. I. Kim, A.J. Barnes, J.M. Oyler, R. Schepers, R.E. Joseph, Jr., E.J. Cone, D. Lafko, E.T. Moolchan, and M.A. Huestis. Plasma and oral fluid pharmacokinetics and pharmacodynamics after oral codeine administration. *Clin. Chem.* 48: 1486–1496 (2002).

178. R.J. Schepers, J.M. Oyler, R.E. Joseph, Jr., E.J. Cone, E.T. Moolchan, and M.A. Huestis. Methamphetamine and amphetamine pharmacokinetics in oral fluid and plasma after controlled oral methamphetamine administration to human volunteers. *Clin. Chem.* 49: 121–132 (2003).

179. E.J. Cone, P. Welch, B.D. Paul, and J.M. Mitchell. Forensic drug testing for opiates. III. Urinary excretion rates of morphine and codeine following codeine administration. *J. Anal. Toxicol.* 15: 161–166 (1991).

180. H.N. ElSohly, D.F. Stanford, A.B. Jones, M.A. ElSohly, H. Snyder, and C. Pedersen. Gas chromatographic/mass spectrometric analysis of morphine and codeine in human urine of poppy seed eaters. *J. Forensic Sci.* 33: 347–356 (1988).

181. P. Lafolie, O. Beck, Z. Lin, F. Albertioni, and L. Boreus. Urine and plasma pharmacokinetics of codeine in healthy volunteers: implications for drugs-of-abuse testing. *J. Anal. Toxicol.* 20: 541–546 (1996).

182. F. Mari and E. Bertol. Observations on urinary excretion of codeine in illicit heroin addicts. *J. Pharm. Pharmacol.* 33: 814–815 (1981).

183. J.M. Oyler, E.J. Cone, R.E. Joseph, Jr., and M.A. Huestis. Identification of hydrocodone in human urine following controlled codeine administration. *J. Anal. Toxicol.* 24: 530–535 (2000).

184. G. Ceder and A.W. Jones. Concentrations of unconjugated morphine, codeine and 6-acetylmorphine in urine specimens from suspected drugged drivers. *J. Forensic Sci.* 47(2): 366–368 (2002).

185. M.E. Sharp, S.M. Wallace, K.W. Hindmarsh, and H.W. Peel. Monitoring saliva concentrations of methaqualone, codeine, secobarbital, diphenhydramine and diazepam after single oral doses. *J. Anal. Toxicol.* 7: 11–14 (1983).

186. G.J. Di-Gregorio, A.J. Piraino, B.T. Nagle, and E.K. Knaiz. Secretion of drugs by the parotid glands of rats and human beings. *J. Dental Res.* 56: 502–509 (1977).

187. C.L. O'Neal, D.J. Crouch, D.E. Rollins, and A.A. Fatah. The effects of collection methods on oral fluid codeine concentrations. *J. Anal. Toxicol.* 24: 536–542 (2000).

188. G. Skopp, L. Potsch, K. Klinder, B. Richter, R. Aderjan, and R. Mattern. Saliva testing after single and chronic administration of dihydrocodeine. *Int. J. Legal Med.* 114: 133–140 (2001).

189. P. Kintz, A. Tracqui, C. Jamey, and P. Mangin. Detection of codeine and phenobarbital in sweat collected with a sweat patch. *J. Anal. Toxicol.* 20: 197–201 (1996).

190. A. Tracqui, P. Kintz, B. Ludes, C. Jamey, and P. Mangin. The detection of opiate drugs in nontraditional specimens (clothing): a report of ten cases. *J. Forensic Sci.* 40: 263–265 (1995).

191. D.E. Rollins, D.G. Wilkins, G.G. Krueger, M.P. Augsberger, A. Mizuno, C. O'Neal, C.R. Borges, and M.H. Slawson. The effect of hair color on the incorporation of codeine into human hair. *J. Anal. Toxicol.* 27: 545–551 (1999).

192. V. Cirimele, P. Kintz, and P. Mangin. Drug concentrations in human hair after bleaching. *J. Anal. Toxicol.* 19: 331 (1995).

193. S.P. Gygi, R.E. Joseph, Jr., E.J. Cone, D.G. Wilkins, and D.E. Rollins. Incorporation of codeine and metabolites into hair. Role of pigmentation. *Drug Metab. Dispos.* 24: 495–501 (1996).

194. R. Kronstrand, S. Forstberg-Peterson, B. Kagedal, J. Ahlner, and G. Larson. Codeine concentration in hair after oral administration is dependent on melanin content. *Clin. Chem.* 45: 1485–1494 (1999).

195. H. Sachs, R. Denk, and I. Raff. Determination of dihydrocodeine in hair of opiate addicts by GC/MS. *Int. J. Legal. Med.* 105: 247–250 (1993).

196. M. Scheller and H. Sachs. The detection of codeine abuse by hair analysis. *Dtsch. Med. Wochenschr.* 115: 1313–1315 (1990).

197. D.G. Wilkins, H.M. Haughey, G.G. Krueger, and D.E. Rollins. Disposition of codeine in female human hair after multiple-dose administration. *J. Anal. Toxicol.* 19: 492–498 (1995).

198. D.N. Bailey and J.J. Guba. Gas-chromatographic analysis for phencyclidine in plasma, with use of a nitrogen detector. *Clin. Chem.* 26: 437–440 (1980).

199. D.N. Bailey. Phencyclidine abuse. Clinical findings and concentrations in biological fluids after nonfatal intoxication. *Am. J. Clin. Pathol.* 72: 795–799 (1979).

200. D.N. Bailey, R.F. Shaw, and J.J. Guba. Phencyclidine abuse: plasma levels and clinical findings in casual users and in phencyclidine-related deaths. *J. Anal. Toxicol.* 2: 233–237 (1978).

201. C.E. Cook, D.R. Brine, A.R. Jeffcoat, J.M. Hill, M.E. Wall, M. Perez-Reyes, and S.R. DiGuiseppi. Phencyclidine disposition after intravenous and oral doses. *Clin. Pharmacol. Ther.* 31: 625–634 (1982).

202. E.F. Domino and A.E. Wilson. Effects of urine acidification on plasma and urine phencyclidine levels in overdosage. *Clin. Pharmacol. Ther.* 22: 421–424 (1977).

203. J.O. Donaldson and R.C. Baselt. CSF phencyclidine. *Am. J. Psychiatry* 136: 1341–1342 (1979).

204. S. Kerrigan and J.W. Phillips, Jr. Comparison of ELISAs for opiates, methamphetamine, cocaine metabolite, benzodiazepines, phencyclidine, and cannabinoids in whole blood and urine. *Clin. Chem.* 47: 540–547 (2001).

205. G.W. Kunsman, B. Levine, A. Costantino, and M.L. Smith. Phencyclidine blood concentrations in DRE cases. *J. Anal. Toxicol.* 21: 498–502 (1997).

206. A.L. Misra, R.B. Pontani, and J. Bartolomeo. Persistence of phencyclidine (PCP) and metabolites in brain and adipose tissue and implications for long-lasting behavioural effects. *Res. Commun. Chem. Path. Pharmacol.* 24: 431–445 (1979).

207. A.E. Wilson and E.F. Domino. Plasma phencyclidine pharmacokinetics in dog and monkey using a gas chromatography selected ion monitoring assay. *Biomed. Mass Spectrom.* 5: 112–116 (1978).

208. D.N. Bailey. Percutaneous absorption of phencyclidine hydrochloride in vivo. *Res. Commun. Subst. Abuse* 1: 443–450 (1980).

209. W. Bronner, P. Nyman, and D. von Minden. Detectability of phencyclidine and 11-nor-delta-9-tetrahydrocannabinol-9-carboxylic acid in adulterated urine by radioimmunoassay and fluorescence polarization immunoassay. *J. Anal. Toxicol.* 14: 368–371 (1990).

210. E.J. Cone, D.B. Vaupel, and D. Yousefnejad. Monohydroxymetabolites of phencyclidine (PCP): activities and urinary excretion by rat, dog, and mouse. *J. Pharm. Pharmacol.* 34: 197–199 (1982).

211. C.C. Stevenson, D.L. Cibull, G.E. Platoff, D.M. Bush, and J.A. Gere. Solid phase extraction of phencyclidine from urine followed by capillary gas chromatography/mass spectrometry. *J. Anal. Toxicol.* 16: 337–339 (1992).

212. D.N. Bailey and J.J. Guba. Measurement of phencyclidine in saliva. *J. Anal. Toxicol.* 4: 311–313 (1980).

213. M.M. McCarron, C.B. Walberg, J.R. Soares, S.J. Gross, and R.C. Baselt. Detection of phencyclidine usage by radioimmunoassay of saliva. *J. Anal. Toxicol.* 8: 197–201 (1984).

214. W.A. Baumgartner, V.A. Hill, and W.H. Blahd. Hair analysis for drugs of abuse. *J. Forensic Sci.* 34: 1433–1453 (1989).

215. T. Sakamoto and A. Tanaka. Determination of PCP and its major metabolites, PCHP and PPC, in rat hair after administration of PCP. In *Proceedings of the 1994 JOINT TIAFT/SOFT International Meeting*, V. Spiehler, Ed. TIAFT/SOFT Joint Congress, 1994, 215–224.

216. D.A. Kidwell. Analysis of phencyclidine and cocaine in human hair by tandem mass spectrometry. *J. Forensic Sci.* 38: 272–284 (1993).

217. Y. Nakahara, K. Takahashi, T. Sakamoto, A. Tanaka, V.A. Hill, and W.A. Baumgartner. Hair analysis for drugs of abuse. XVII. Simultaneous detection of PCP, PCHP, and PCPdiol in human hair for confirmation of PCP use. *J. Anal. Toxicol.* 21: 356–362 (1997).

218. T. Sakamoto, A. Tanaka, and Y. Nakahara. Hair analysis for drugs of abuse. XII. Determination of PCP and its major metabolites, PCHP and PPC, in rat hair after administration of PCP. *J. Anal. Toxicol.* 20: 124–130 (1996).

219. Y. Nakahara, R. Kikura, T. Sakamoto, T. Mieczkowski, F. Tagliaro, and R. L. Foltz. Findings in hair analysis for some hallucinogens (LSD, MDA/MDMA and PCP). In *Hair Analysis in Forensic Toxicology: Proceedings of the 1995 International Conference and Workshop*, R.A. de Zeeuw, I. Al Hosani, and S. Al Munthiri, Eds. Organizing Committee of the Conference, Abu Dhabi, 1995, 161–184.

220. H. Gjerde, I. Hasvold, G. Pettersen, and A.S. Christophersen. Determination of amphetamine and methamphetamine in blood by derivatization with perfluorooctanoyl chloride and gas chromatography/mass spectrometry. *J. Anal. Toxicol.* 17: 65–68 (1993).

221. S. Cheung, H. Nolte, S.V. Otton, R.F. Tyndale, P.H. Wu, and E.M. Sellers. Simultaneous gas chromatographic determination of methamphetamine, amphetamine and their *p*-hydroxylated metabolites in plasma and urine. *J. Chromatogr.* 690: 77–87 (1997).

222. Y. Gaillard, F. Vayssette, and G. Pepin. Compared interest between hair analysis and urinalysis in doping controls. Results for amphetamines, corticosteroids and anabolic steroids in racing cyclists. *Forensic Sci. Int.* 107: 361–379 (2000).

223. S.B. Matin, S.H. Wan, and J.B. Knight. Quantitative determination of enantiomeric compounds. I. Simultaneous measurement of the optical isomers of amphetamine in human plasma and saliva using chemical ionization mass spectrometry. *Biomed. Mass Spectrom.* 4: 118–121 (1977).

224. F.P. Smith. Detection of amphetamine in bloodstains, semen, seminal stains, saliva, and saliva stains. *Forensic Sci. Int.* 17: 225–228 (1981).

225. J.M. Davis, I.J. Kopin, L. Lemberger, and J. Axelrod. Effects of urinary pH on amphetamine metabolism. *Ann. N. Y. Acad. Sci.* 179: 493–501 (1971).

226. B.D. Paul, M.R. Past, R.M. McKinley, J.D. Foreman, L.K. McWhorter, and J.J. Snyder. Amphetamine as an artifact of methamphetamine during periodate degradation of interfering ephedrine, pseudoephedrine, and phenylpropanolamine: an improved procedure for accurate quantitation of amphetamines in urine. *J. Anal. Toxicol.* 18: 331–336 (1994).

227. G.A. Alles and B.B. Wisegarver. Amphetamine excretion studies in man. *Toxicol. App. Pharmacol.* 3: 678–688 (1961).

228. A.H. Beckett and M. Rowland. Determination and identification of amphetamine in urine. *J. Pharm. Pharmacol.* 17: 59–60 (1965).

229. J. D'Nicuola, R. Jones, B. Levine, and M.L. Smith. Evaluation of six commercial amphetamine and methamphetamine immunoassays for cross-reactivity to phenylpropanolamine and ephedrine in urine. *J. Anal. Toxicol.* 16: 211–213 (1992).

230. A.H. Beckett and M. Rowland. Urinary excretion kinetics of amphetamine in man. *J. Pharm. Pharmacol.* 17: 628–638 (1965).

231. D.E. Blandford and P.R.E. Desjardins. Detection and identification of amphetamine and methamphetamine in urine by GC/MS. *Clin. Chem.* 40: 145–147 (1994).

232. J.T. Cody and S. Valtier. Detection of amphetamine and methamphetamine following administration of benzphetamine. *J. Anal. Toxicol.* 22: 299–309 (1998).

233. B.K. Gan, D. Baugh, R.H. Liu, and A.S. Walia. Simultaneous analysis of amphetamine, methamphetamine, and 3,4-methylenedioxymethamphetamine MDMA in urine samples by solid-phase extraction, derivatization, and gas chromatography/mass spectrometry. *J. Forensic Sci.* 36: 1331–1341 (1991).

234. L.M. Gunne. The urinary output of d-and l-amphetamine in man. *Biochem. Pharmacol.* 16: 863–869 (1967).

235. S.J. Mule and G.A. Casella. Confirmation of marijuana, cocaine, morphine, codeine, amphetamine, methamphetamine, phencyclidine by GC/MS in urine following immunoassay screening. *J. Anal. Toxicol.* 12: 102–107 (1988).

236. A. Nice and A. Maturen. False-positive urine amphetamine screen with Ritodrine. *Clin. Chem.* 35: 1542–1543 (1989).

237. J.L. Valentine, G.L. Kearns, C. Sparks, L.G. Letzig, C.R. Valentine, S.A. Shappell, D.F. Neri, and C.A. DeJohn. GC/MS determination of amphetamine and methamphetamine in human urine for 12 hours following oral administration of dextro-methamphetamine: lack of evidence supporting the established forensic guidelines for methamphetamine confirmation. *J. Anal. Toxicol.* 19: 581–590 (1995).

238. M.R. Peace, L.D. Tarnai, and A. Poklis. Performance evaluation of four on-site drug-testing devices for detection of drugs of abuse in urine. *J. Anal. Toxicol.* 24: 589–594 (2000).

239. A. Poklis and K.A. Moore. Response of EMIT amphetamine immunoassays to urinary desoxyephedrine following Vicks inhaler use. *Ther. Drug Monit.* 17: 89–94 (1995).

240. A. Smith-Kielland, B. Skuterud, and J. Morland. Urinary excretion of amphetamine after termination of drug abuse. *J. Anal. Toxicol.* 21: 325–329 (1997).

241. H. Vapaatalo, S. Karkkainen, and K.E.O. Senius. Comparison of saliva and urine samples in thin-layer chromatographic detection of central nervous stimulants. *Int. J. Clin. Pharmacol.* 1: 5–8 (1984).

242. E.M. Thurman, M.J. Pedersen, R.L. Stout, and T. Martin. Distinguishing sympathomimetic amines from amphetamine and methamphetamine in urine by gas chromatography/mass spectrometry. *J. Anal. Toxicol.* 16: 19–27 (1992).

243. K. Takahashi. Determination of methamphetamine and amphetamine in biological fluids and hair by gas chromatography. *Dep. Legal Med.* 38: 319–336 (1984).

244. S.H. Wan, S.B. Matin, and D.L. Azarnoff. Kinetics, salivary excretion of amphetamine isomers, and effect of urinary pH. *Clin. Pharmacol. Ther.* 23: 585–590 (1978).

245. T.B. Vree, A.T. Muskens, and J.M. Van Rossum. Excretion of amphetamines in human sweat. *Arch. Int. Pharmacodyn. Ther.* 199: 311–317 (1972).

246. Y. Nakahara and R. Kikura. Hair analysis for drugs of abuse. XIII. Effect of structural factors on incorporation of drugs into hair: the incorporation rates of amphetamine analogs. *Arch. Toxicol.* 70: 841–849 (1996).

247. M. Rothe, F. Pragst, K. Spiegel, T. Harrach, K. Fischer, and J. Kunkel. Hair concentrations and self-reported abuse history of 20 amphetamine and ecstasy users. *Forensic Sci. Int.* 89: 111–128 (1997).

248. F. Musshoff, H.P. Junker, D.W. Lachenmeier, L. Kroener, and B. Madea. Fully automated determination of amphetamines and synthetic designer drugs in hair samples using headspace solid-phase microextraction and gas chromatography-mass spectrometry. *J. Chromatogr. Sci.* 40: 359–364 (2002).

249. F. Musshoff, D.W. Lachenmeier, L. Kroener, and B. Madea. Automated headspace solid-phase dynamic extraction for the determination of amphetamines and synthetic designer drugs in hair samples. *J. Chromatogr. A* 958: 231–238 (2002).

250. P. Kintz and V. Cirimele. Interlaboratory comparison of quantitative determination of amphetamine and related compounds in hair samples. *Forensic Sci. Int.* 84: 151–156 (1997).

251. D.L. Allen and J.S. Oliver. The use of supercritical fluid extraction for the determination of amphetamines in hair. *Forensic Sci. Int.* 107: 191–199 (2000).

252. P. Kintz, V. Cirimele, A. Tracqui, and P. Mangin. Simultaneous determination of amphetamine, methamphetamine, 3,4-methylenedioxyamphetamine and 3,4-methylenedioxymethamphetamine in human hair by gas chromatography-mass spectrometry. *J. Chromatogr.* 670: 162–166 (1995).

253. J. Rohrich and G. Kauert. Determination of amphetamine and methylenedioxy-amphetamine-derivatives in hair. *Forensic Sci. Int.* 84: 179–188 (1997).

254. O. Suzuki, H. Hattori, and M. Asano. Detection of methamphetamine and amphetamine in a single human hair by gas chromatography/chemical ionization mass spectrometry. *J. Forensic Sci.* 29: 611–617 (1984).

255. R. Kikura and Y. Nakahara. Distinction between amphetamine-like OTC drug use and illegal amphetamine/methamphetamine use by hair analysis. In *Proceedings of the 1994 JOINT TIAFT/SOFT International Meeting*, V. Spiehler, Ed. TIAFT/SOFT, 1994, 236–247.

256. B.K. Logan, C.L. Fligner, and T. Haddix. Cause and manner of death in fatalities involving methamphetamine. *J. Forensic Sci.* 43(1): 28–34 (1998).

257. B.K. Logan. Methamphetamine and driving impairment. *J. Forensic Sci.* 41: 457–464 (1996).

258. S. Rasmussen, R. Cole, and V. Spiehler. Methamphetamine in antemortem blood and urine by radioimmunoassay and GC/MS. *J. Anal. Toxicol.* 13: 263–267 (1989).

259. V.R. Spiehler, I.B. Collison, P.R. Sedgwick, S.L. Perez, S.D. Le, and D.A. Farnin. Validation of an automated microplate enzyme immunoassay for screening of postmortem blood for drugs of abuse. *J. Anal. Toxicol.* 22: 573–579 (1998).

260. R.C. Driscoll, F.S. Barr, B.J. Gragg, and G.W. Moore. Determination of therapeutic blood levels of methamphetamine and pentobarbital by GC. *J. Pharm. Sci.* 60(10): 1492–1495 (1971).

261. R.L. Fitzgerald, J.M. Ramos, S.C. Bogema, and A. Poklis. Resolution of methamphetamine stereoisomers in urine drug testing: urinary excretion of R(–)-methamphetamine following use of nasal inhalers. *J. Anal. Toxicol.* 12: 255–259 (1988).

262. J.T. Cody. Determination of methamphetamine enantiomer ratios in urine by gas chromatography-mass spectrometry. *J. Chromatogr.* 580: 77–95 (1992).

263. J. Sukbuntherng, A. Hutchaleelaha, H.H. Chow, and M. Mayersohn. Separation and quantitation of the enantiomers of methamphetamine and its metabolites in urine by HPLC: precolumn derivatization and fluorescence detection. *J. Anal. Toxicol.* 19: 139–147 (1995).

264. C.L. Hornbeck, J.E. Carrig, and R.J. Czarny. Detection of a GC/MS artifact peak as methamphetamine. *J. Anal. Toxicol.* 17: 257–263 (1993).

265. M.R. Lee, S.C. Yu, C.L. Lin, Y.C. Yeh, Y.L. Chen, and S.H. Hu. Solid-phase extraction in amphetamine and methamphetamine analysis of urine. *J. Anal. Toxicol.* 21: 278–282 (1997).

266. K. McCambly, R.C. Kelly, T. Johnson, J.E. Johnson, and W.C. Brown. Robotic solid-phase extraction of amphetamines from urine for analysis by gas chromatography-mass spectrometry. *J. Anal. Toxicol.* 21: 438–444 (1997).

267. A. Poklis and K.A. Moore. Stereoselectivity of the TDxADx/FLx amphetamine/methamphetamine II amphetamine/methamphetamine immunoassays — response of urine specimens following nasal inhaler use. *Clin. Toxicol.* 33: 35–41 (1995).

268. J.M. Oyler, E.J. Cone, R.E. Joseph, Jr., E.T. Moolchan, and M.A. Huestis. Duration of detectable methamphetamine and amphetamine excretion in urine after controlled oral administration of methamphetamine to humans. *Clin. Chem.* 48: 1703–1714 (2002).

269. S. Suzuki, T. Inoue, H. Hori, and S. Inayama. Analysis of methamphetamine in hair, nail, sweat, and saliva by mass fragmentography. *J. Anal. Toxicol.* 13: 176–178 (1989).

270. J. Fay, R. Fogerson, D. Schoendorfer, R.S. Niedbala, and V. Spiehler. Detection of methamphetamine in sweat by EIA and GC-MS. *J. Anal. Toxicol.* 20: 398–403 (1996).

271. Y.C. Yoo, H.S. Chung, H.K. Choi, and W.K. Jin. Determination of methamphetamine in the hair of Korean drug abusers by GC/MS. In *Proceedings of the 1994 JOINT TIAFT/SOFT International Meeting*, V. Spiehler, Ed. TIAFT/SOFT Joint Congress, 1994, 207–214.

272. I. Ishiyama, T. Nagai, and S. Toshida. Detection of basic drugs (methamphetamine, antidepressants, and nicotine) from human hair. *J. Forensic Sci.* 28: 380–385 (1983).

273. R. Kikura and Y. Nakahara. Hair analysis for drugs of abuse. IX. Comparison of deprenyl use and methamphetamine use by hair analysis. *Biol. Pharm. Bull.* 18: 267–272 (1995).

274. Y. Nakahara, R. Kikura, M. Yasuhara, and T. Mukai. Hair analysis for drug abuse. XIV. Identification of substances causing acute poisoning using hair root. I. Methamphetamine. *Forensic Sci. Int.* 84: 157–164 (1997).

275. S. Suzuki, T. Inoue, T. Yasuda, T. Niwaguchi, H. Hori, and S. Inayama. Analysis of methamphetamine in human hair by fragmentography. *Eisei Kagaku* 30: 23–26 (1984).

276. M.R. Moeller and M. Hartung. Ecstasy and related substances — serum levels in impaired drivers. *J. Anal. Toxicol.* 21: 501–501 (1997).

277. M. Navarro, S. Pichini, M. Farre, J. Ortuno, P.N. Roset, J. Segura, and R. de la Torre. Usefulness of saliva for measurement of 3,4-methylenedioxymethamphetamine and its metabolites: correlation with plasma drug concentrations and effect of salivary pH. *Clin. Chem.* 47: 1788–1795 (2001).

278. T.P. Rohrig and R.W. Prouty. Tissue distribution of methylenedioxymethamphetamine. *J. Anal. Toxicol.* 16: 52–53 (1992).

279. R. de la Torre, M. Farre, J. Ortuno, M. Mas, R. Brenneisen, P.N. Roset, J. Segura, and J. Cami. Non-linear pharmacokinetics of MDMA ("ecstasy") in humans. *Br. J. Clin. Pharmacol.* 49: 104–109 (2000).

280. T. Kraemer and H.H. Maurer. Toxicokinetics of amphetamines: metabolism and toxicokinetic data of designer drugs, amphetamine, methamphetamine, and their *N*-alkyl derivatives. *Ther. Drug Monit.* 24: 277–289 (2002).

281. N. Samyn, G. De Boeck, M. Wood, C.T. Lamers, D. De Waard, K.A. Brookhuis, A.G. Verstraete, and W.J. Riedel. Plasma, oral fluid and sweat wipe ecstasy concentrations in controlled and real life conditions. *Forensic Sci. Int.* 128: 90–97 (2002).

282. A. Poklis, R.L. Fitzgerald, K.V. Hall, and J.J. Saady. EMIT-d.a.u. monoclonal amphetamine/methamphetamine assay. II. Detection of methylenedioxyamphetamine (MDA) and methylenedioxymethamphetamine (MDMA). *Forensic Sci. Int.* 59: 63–70 (1993).

283. G.W. Kunsman, B. Levine, J.J. Kuhlman, R.L. Jones, R.O. Hughes, C.I. Fujiyama, and M.L. Smith. MDA-MDMA concentrations in urine specimens. *J. Anal. Toxicol.* 20: 517–521 (1996).

284. J.M. Ramos, Jr., R.L. Fitzgerald, and A. Poklis. MDMA and MDA cross reactivity observed with Abbott TDx amphetamine/methamphetamine reagents. *Clin. Chem.* 34: 991 (1988).

285. H.H. Maurer, J. Bickeboeller-Friedrich, T. Kraemer, and F.T. Peters. Toxicokinetics and analytical toxicology of amphetamine-derived designer drugs ("Ecstasy"). *Toxicol. Lett.* 112–113: 133–142 (2000).

286. R. Pacifici, M. Farre, S. Pichini, J. Ortuno, P.N. Roset, P. Zuccaro, J. Segura, and R. de la Torre. Sweat testing of MDMA with the Drugwipe analytical device: a controlled study with two volunteers. *J. Anal. Toxicol.* 25: 144–146 (2001).

287. S. Pichini, M. Navarro, R. Pacifici, P. Zuccaro, J. Ortuno, M. Farre, P.N. Roset, J. Segura, and R. de la Torre. Usefulness of sweat testing for the detection of MDMA after a single-dose administration. *J. Anal. Toxicol.* 27: 294–303 (2003).

288. F. Tagliaro, Z. De Battisti, A. Groppi, Y. Nakahara, D. Scarcella, R. Valentini, and M. Marigo. High sensitivity simultaneous determination in hair of the major constituents of ecstasy (3,4-methylene-dioxymethamphetamine, 3,4-methylenedioxyamphetamine and 3,4-methylene-dioxyethylamphetamine) by high-performance liquid chromatography with direct fluorescence detection. *J. Chromatogr. B Biomed. Sci. Appl.* 723: 195–202 (1999).

289. M. Uhl. Tandem mass spectrometry: a helpful tool in hair analysis for the forensic expert. *Forensic Sci. Int.* 107: 169–179 (2000).

290. Y. Nakahara and R. Kikura. Hair analysis for drugs of abuse. XVIII. 3,4-Methylenedioxymethamphetamine (MDMA) disposition in hair roots and use in identification of acute MDMA poisoning. *Biol. Pharm. Bull.* 20: 969–972 (1997).

291. C. Girod and C. Staub. Analysis of drugs of abuse in hair by automated solid-phase extraction, GC/EI/MS and GC ion trap/CI/MS. *Forensic Sci. Int.* 107: 261–271 (2000).

292. R. Kikura, Y. Nakahara, T. Mieczkowski, and F. Tagliaro. Hair analysis for drug abuse. XV. Disposition of 3,4-methylenedioxymethamphetamine (MDMA) and its related compounds into rat hair and application to hair analysis for MDMA abuse. *Forensic Sci. Int.* 84: 165–177 (1997).

CHAPTER 9

Specimen Validity Testing

Yale H. Caplan, Ph.D., DABFT
National Scientific Services, Baltimore, Maryland

CONTENTS

Urine drug testing is, by its nature and the privacy considerations, decisions, and laws of the federal government and the states, an unobserved process. This is the standard practice unless there is specific individual suspicion that a specimen has been altered or substituted. Although the practice of tampering had been reported prior to the implementation of the federally regulated workplace drug-testing program, the problem became evident in the 1990s with the introduction of "Urine Aid" and later "Klear," products among others advertised to "Beat a Drug Test." Earlier methods were crude and often ineffective utilizing more commonly available products such as salt, bleach, soap, and vinegar and sharing information about such products by "word of mouth." Later a significant cottage industry grew through health food stores, advertisements in *High Times* and other magazines, and extensive citation on the Internet.

How effective can a drug-testing program be if a magic potion priced at $20 to $30 can provide a person a means to evade detection? In a deterrent-based program, credibility can rapidly decline if users believe they can beat the test. The Department of Health and Human Services (HHS) and the Department of Transportation (DOT) developed countermeasures by causing laboratories to inspect and test specimens, verify their normal composition, and search for foreign chemicals. The term *specimen validity testing* (SVT) was coined to identify a sequence of testing designed to check if urine was really urine and if the urine was normal or adulterated. This task was not so simple. Laboratories developed credible tests for the five classes of drugs, substances of known composition and predictable outcome, and were now faced with developing tests for a variety of specimen characteristics and an array of unknown compounds. Indeed, the task of comprehensively searching for normal elements that comprise urine and ascertaining that no foreign materials have been added can be more complex and costly than the drug testing itself.

The intent of the donor is to subvert the drug testing procedure and create a false-negative result. This can be accomplished in three ways: (1) by diluting the urine through excessive fluid intake, (2) by substituting other urine, hopefully drug free urine, in place of the donor's urine, and (3) adulterating the urine by adding a chemical to the specimen that either destroys the drug in the urine or otherwise interferes with the laboratory immunoassay tests.

9.1 CHARACTERISTICS OF URINE

Urine is an aqueous solution produced by the kidneys. A review of urine characterization with emphasis on workplace testing has been compiled.[1] Urine's major constituents are primarily electrolytes, metabolic excretory products, and other substances eliminated through the kidneys. The initial characterization of a urine specimen is based on its appearance. Color, clarity, odor, and foaming properties contribute to the appearance of urine.

The color of a urine specimen is related to the concentration of its various constituents, most notably urochromes, which exhibit yellow, brown, and red pigments. A normal first morning void has a distinct deep yellow color. It should not be colorless. The characteristic yellow color is predominately caused by the presence of urobilinogen, a hemoglobin breakdown product. After hydration, urine is usually straw-colored, indicating dilute urine. Very dilute urine is essentially colorless and has a water-like appearance.

A normal fresh void is clear and transparent. Freshly voided urine that is cloudy or turbid can indicate the presence of white blood cells, red blood cells, epithelial cells, or bacteria. Upon standing, flaky precipitates from urinary tract mucin may appear in the specimen. Aged alkaline urine may become cloudy because of crystal precipitation. After a lipid-rich meal, urine may also become alkaline and cloudy.

Freshly voided urine is normally odorless. With age, urine acquires a characteristic aromatic odor. As the constituents in the urine decompose, ammonia, putrefaction compounds, and hydrogen sulfide are detected. Certain foodstuffs, such as coffee, garlic, or asparagus, give a distinctive scent. Patients with poorly controlled diabetes produce ketone bodies such as acetone, which impart a fruity odor to urine.

Urine foaming is caused by the presence of protein in the specimen, and foam bubbles should not exhibit the rainbow appearance that is indicative of soap contamination.

9.2 CHEMICAL CONSTITUENTS AND OTHER CHARACTERISTICS OF URINE

The kidneys filter plasma, reabsorb most of the dissolved substances that are filtered, secrete some of these substances back into its filtrate, and leave behind a concentrated solution of metabolic waste known as urine. Metabolic waste products are present in urine at high concentrations.

Creatinine is spontaneously and irreversibly formed from creatine and creatine phosphate in muscle. Creatinine, creatine's anhydride, is a metabolic waste product that is not reutilized by the body. Because there is a direct relationship between creatinine formation and muscle mass, creatinine production is considered constant from day to day provided that muscle mass remains unchanged. In view of this constant production, random creatinine results approximate 24-h collection reference intervals. Creatinine is freely filtered by the renal glomeruli and is not significantly reabsorbed by the renal tubules, but a small amount is excreted by active renal tubular secretion. Creatinine is excreted at a relatively constant rate relatively independent of diuresis, provided kidney function is not impaired. Creatinine production and excretion are age and sex dependent. Normal urine creatinine concentrations are greater than 20 mg/dl. Abnormal levels of creatinine may result from excessive fluid intake, glomerulonephritis, pyleonephritis, reduced renal blood flow, renal failure, myasthenia gravis, or a high meat diet.

Specific gravity assesses urine concentration, or amount of dissolved substances present in a solution. As increasing amounts of substances are added to urine, the concentration of these dissolved substances and the density, or the weight of the dissolved substances per unit volume of liquid, increase. Specific gravity varies greatly with fluid intake and state of hydration. Normal values for the specific gravity of human urine range from approximately 1.0020 to 1.0200. Decreased urine specific gravity values may indicate excessive fluid intake, renal failure, glomerulonephritis, pyelonephritis, or diabetes insipidus. Increased urine specific gravity values may result from dehydration, diarrhea, excessive sweating, glucosuria, heart failure, proteinuria, renal arterial stenosis, vomiting, and water restriction.

pH is the inverse logarithmic function of the hydrogen ion concentration. It serves as an indicator of the acidity of a solution. The two organs that are primarily responsible for regulating the extremely narrow blood pH range that is compatible with human life are the lungs and the kidneys. The kidneys maintain the blood pH range by eliminating metabolic waste products. The pH of the urine is used clinically to assess the ability of the kidneys to eliminate toxic substances. Urinary pH undergoes physiological fluctuations throughout the day. Urinary pH values are decreased in the early morning followed by an increase in the late morning and early afternoon. In the bacteria-contaminated urine specimen, pH will increase upon standing because of bacterial ammonia formation. Normal urinary pH values range from 4.5 to 9.0.

9.3 THE ROLE OF HHS AND DOT IN SPECIMEN VALIDITY TESTING

HHS through its Mandatory Guidelines for Federal Workplace Drug Testing and other notices has directed laboratories in the conduct of specimen validity tests. The early guidelines permitted testing but did not define the characteristics. When the adulteration and substitution issues became more prominent, HHS and DOT issued guidance. These were considered voluntary.

HHS–SAMHSA Program Document (PD) 33
   Title: Testing Split (Bottle B) Specimen for Adulterants
   Dated: March 9, 1998
   Summary: The guidance established technical threshold "cutoff" values for pH and nitrates. It required laboratory testing for pH and nitrite concentration in any split specimen that failed to reconfirm. It also authorized additional adulteration testing for the presence of glutaraldehyde, surfactants, bleach, and other adulterants if indicated.
HHS–SAMHSA Program Document (PD) 35
   Title: Guidelines for Reporting Specimen Validity Test Results
   Dated: September 28, 1998
   Summary: PD 35 replaced the Category I, II, and III reporting protocols established in 1993 with more detailed laboratory test reporting protocols. It established unacceptable limits for nitrite concentration, pH, and substitution. It required further testing when a split specimen failed to reconfirm.

U.S. DOT–Office of Drug and Alcohol Policy and Compliance (ODAPC)
    Title: MRO Guidance for Interpreting Specimen Validity Test Results
    Dated: September 28, 1998
    Summary: This companion document to PD 35 contained formal regulatory guidance for MROs,
        detailing how they were to respond to the various laboratory reporting protocols.
HHS–SAMHSA Program Document (PD) 37
    Title: Specimen Validity Testing
    Dated: July 28, 1999
    Summary: PD 37 provided specific technical guidelines to the laboratories. The document notes that
        specimen validity testing **may** be conducted on Bottle A and *must* be conducted on Bottle B if
        Bottle B fails to reconfirm for the requested drug/analyte. Specimen validity tests *may* include,
        but are not limited to, tests for creatinine concentration, specific gravity, pH, nitrite concentra-
        tion, pyridine, glutaraldehyde, bleach, and soap. These tests must be performed using methods
        that are validated by the laboratory.

Subsequently, DOT published its final rule, 49CFR Part 40, on December 19, 2000[2] and a technical amendment to the Final Rule on August 1, 2001.[3] The 2000 rule mandated validity testing but the 2001 amendment changed this to *authorized* but *not mandated* pending future HHS actions. The rule section 40.91 states:
What validity tests must laboratories conduct on primary specimens?

Creatinine and specific gravity (SG)
SG if creatinine < 20 mg/dl
Measure pH
Substances that may be used to adulterate urine
Send to second laboratory if unable to confirm adulterant
New adulterant, report to DOT and HHS
Complete testing for drugs
Conserve specimen

Most recently HHS published its Revised Mandatory Guidelines for Workplace Drug Testing Programs on April 13, 2004 with an effective date of November 1, 2004.[4] It defined the final SVT requirements. Notably it included the following:

1. Creatinine concentration criterion defining a substituted specimen changed to <2 mg/dl
2. SG analysis with four decimal places required to report as substituted or as invalid based on creatinine and SG
3. Added definitions specifically associated with SVT
   - Initial drug test, initial validity test
   - Confirmatory drug test, confirmatory validity test
   - Dilute specimen, adulterated specimen
   - Donor empties pockets and displays for collector
4. Included all the reporting requirements to report a specimen adulterated, substituted, diluted, or as an invalid result
5. Added a requirement to report the actual numerical values (concentrations) for adulterated results, and the confirmatory creatinine concentration and confirmatory SG result for a substituted specimen
6. Added specimen retention requirements for the laboratory
7. Provided a list of fatal flaws and correctable flaws
8. Expanded retest requirements for drugs
9. Mandated agencies to have BQC that are adulterated or substituted
10. MRO review expanded to include adulterated and substituted specimens and reporting
11. Requires laboratories to conduct drug and validity testing at the same facility
12. Defined requirements for creatinine and SG:
    - Determine the creatinine on every specimen

- Determine the SG if creatinine < 20 mg/dl
- Determine SG to four decimal places if reporting a specimen as substituted or as invalid based on creatinine and SG
13. Determine the pH on every specimen
14. Perform one or more validity tests for oxidizing adulterants on every specimen
15. Perform additional validity tests when the following conditions are observed:
    - Abnormal physical characteristics
    - Reactions or responses characteristic of an adulterant during testing
    - Possible unidentified interfering substance or adulterant
16. May send to second laboratory if unable to confirm adulterant (laboratory and MRO decide)
17. If a new adulterant is identified, report to DOT and HHS
    - Complete testing for drugs
18. Conserve specimen

According to these guidelines:

Substituted is a specimen with
- Creatinine < 2 mg/dl and SG < 1.0010 or
- Creatinine < 2 mg/dl and SG > 1.0200
Dilute is a specimen with
- Creatinine 2 to <20 mg/dl and
- SG 1.0010 to 1.0030
Invalid is a specimen in which inconsistent creatinine concentration and SG results are obtained, i.e.,
- Creatinine less than 2 mg/dl on *both* the initial and confirmatory creatinine tests and SG greater than 1.0010 but less than 1.0200 on *either or both* the initial and confirmatory SG tests
- Creatinine greater than or equal to 2 mg/dl on *either or both* the initial or confirmatory creatinine tests and SG less than or equal to 1.0010 on both the initial and confirmatory SG tests
A urine specimen is reported adulterated when:
- pH < 3
- pH ≥ 11
- Nitrite concentration ≥ 500 µg/ml (two different tests required)
- Chromium (VI) concentration ≥ 50 µg/ml (two different tests required)
- Halogen is detected and confirmed, with a specific concentration ≥ the LOD of the confirmatory test on the second aliquot (two different tests required)
- Glutaraldehyde is detected and confirmed, with a concentration ≥ the LOD of the confirmatory test on the second aliquot (two different tests required)
- Pyridine is detected and confirmed, with a concentration ≥ the LOD of the confirmatory test on the second aliquot (two tests required)
- Surfactant is detected and confirmed, with a ≥100 µg/ml dodecylbenzene sulfonate-equivalent cutoff concentration on the second aliquot (two tests required)
- The presence of any other adulterant not specified is verified using an initial test on the first aliquot and a different confirmatory test on the second aliquot

The reporting requirements are summarized in Table 9.1. The table was provided by HHS as part of the Mandatory Guidelines effective November 1, 2004. It describes the conditions for positive and negative drug test results as well as adulterated and substituted SVT results; however, it expands the number of conditions for which a specimen is reported as invalid. The increased number of SVT tests plus many more tests that cannot be satisfactorily completed caused a significant increase in the number of invalid reports.

9.4 ROLE OF THE MEDICAL REVIEW OFFICER

Medical Review Officers are required to review tests reported as adulterated, substituted, and invalid. Although the HHS Mandatory Guidelines are in effect, some differences exist between

Table 9.1   HHS Reporting Template

Effective 11-1-2004

Result Box(es) checked on Federal CCF (Step 5a) or Result indicated on Electronic Report	Note
Negative	All drug tests are negative and all specimen validity test results are acceptable
Negative and Dilute	For DOT-regulated specimens: provide Creatinine & SpGr numerical values (See Below)
Invalid Result	State reason for invalid result (See Below)
Positive, Specific Drug	No comment required
Positive, Specific Drug, and Dilute	No comment required
Positive, Specific Drug, and Adulterated	State reason for adulterated result (See Below)
Positive, Specific Drug, and Substituted	Provide Creatinine & SpGr numerical values (See Below)
Positive, Specific Drug, and Invalid Result	State reason for invalid result (See Below)
Adulterated	State reason for adulterated result (See Below)
Adulterated and Invalid Result	State reasons for adulterated & invalid results (See Below)
Adulterated and Substituted	State reason for adulterated result & provide Creatinine and SpGr numerical values (See Below)
Substituted	Provide Creatinine and SpGr numerical values (See Below)
Substituted and Invalid Result	Provide Creatinine and SpGr numerical values & state reason for invalid result (See Below)
Rejected for Testing	State reason for rejecting specimen (See Below)

Test Result	Specific Comment Required on Remarks Line (Step 5a) of the CCF and on Electronic Report	Note
Negative and Dilute	For DOT-Regulated specimens: Creatinine = (numerical value) mg/dL & SpGr = (numerical value)	
Substituted	Creatinine = (conf. test numerical value) mg/dL & SpGr = (conf. test numerical value)	
Adulterated	pH = (conf. test numerical value)	
	Nitrite = (conf. test numerical value) mcg/mL	
	Chromium (VI) = (conf. test numerical value) mcg/mL	
	(Specify Halogen) Present	
	Glutaraldehyde = (conf. test numerical value) mcg/mL	
	Pyridine = (conf. test numerical value) mcg/mL	
	Surfactant Present	
	(Specify Adulterant) Present	
Invalid Result	Creatinine < 2 mg/dL; SpGr Acceptable	SpGr >1.0010 & <1.0200
	SpGr ≤ 1.0010; Creatinine ≥ 2 mg/dL	
	Abnormal pH	
	Possible (Characterize as Oxidant, Halogen, Aldehyde, or Surfactant) Activity*	
	Immunoassay Interference*	
	GC/MS interference*	
	Abnormal Physical Characteristic - (Specify)*	
	Bottle A and Bottle B - Different Physical Appearance*	
Rejected for Testing	Fatal Flaw: Specimen ID number mismatch/missing	
	Fatal Flaw: No collector printed name & No signature	
	Fatal Flaw: Tamper-evident seal broken	If redesignation is not possible
	Fatal Flaw: Insufficient specimen volume	
	Uncorrected Flaw: Wrong CCF Used	Wait 5 business days before reporting flaw if uncorrected
	Uncorrected Flaw: Collector signature not recovered	

* Lab shall contact the MRO to discuss the Invalid Result   →   *Revised January 4, 2005*

DOT and HHS requirements. First HHS mandates SVT for the federal agencies it controls. DOT incorporates HHS requirements in its rules but at present they are not entirely compatible. DOT SVT testing is authorized but not mandated. In addition, DOT published an interim final rule on May 28, 2003,[5] changing the creatinine criterion to <2 from 5 mg/dl and an additional interim rule on November 9, 2004[6] to incorporate most of the HHS rule. The differences are as follows:

HHS — Laboratories report quantitative creatinine and SG values for adulterated and substituted only
DOT — Laboratories report quantitative creatinine and SG values for all negative dilute specimens

MROs are required to take the following actions if the creatinine is above 2 mg/dl:

HHS — No requirements
DOT — Creatinine 2 or greater but less than 5 mg/dl with appropriate SG: Recollect under direct observation
DOT — Creatinine 5 or greater but less than 20 mg/dl with appropriate SG: If negative (drug) may, but not required, to take another test immediately

9.5 DILUTION/SUBSTITUTION

A donor may attempt to decrease the concentration of drugs or drug metabolites that may be present in his or her urine by dilution. Deliberate dilution may occur *in vivo* by consuming large volumes of liquid or *in vitro* by adding water or another liquid to the specimen. Donors also have been known to substitute urine specimens with drug-free urine or other liquid during specimen collection. Due to donor privacy considerations, collections for federally regulated drug testing programs are routinely unobserved. Therefore, dilution and substitution may go undetected by collectors and thus provide a viable method for donors to defeat drug tests. There are many products available today purporting to "cleanse" the urine prior to a drug test, some of which are diuretics. Others are designed specifically for urine specimen substitution, including drug-free urine, additives, and containers/devices to aid concealment. Some devices have heating mechanisms to bring the substituted specimen's temperature within the range set by HHS to determine specimen validity at the time of collection (i.e., 32 to 38°C/90 to 100°F). Additional products include prosthetic devices to deceive the collector/observer even during a direct observed collection.

According to federal guidelines, a specimen is:

Dilute if the creatinine is <20 mg/dl and the specific gravity is <1.003, unless the criteria for a substituted specimen are met.

The concentration of drug metabolites in urine is a function of collection time (related to drug use) and relative water content. The goal of the suborner is to dilute the urine and the drugs in it to concentrations that are below the cutoff concentrations for the various drugs. Hydration simply reduces the detection time since last use. It may or may not be significant. Attempts to manage intentional hydration are limited by the fact that normal hydration and specific gravity are highly variable. One cannot unduly restrict water consumption on one hand, and demand an adequate volume of urine from a donor on the other.

There are currently no effective techniques to address the issue of dilution. Most dilute specimens probably occur as the result of hydration — donor drinking large quantities of fluids prior to the collection — rather than the external addition of water to a specimen after it has been provided. Heated water sources should not be available; hence, external hydration should result in detection through temperature checks. Other water sources should be blued with dye to facilitate detection if used.

The most appropriate solution to the dilution problem would be to lower the cutoff values when testing for drugs in dilute specimens. However, this is not allowed in most regulated testing programs (NRC is the exception). The procedures, although technically possible, are not standardized among laboratories in a manner that would assure appropriate uniformity. Such techniques may be valuable, however, in special circumstances for specific situations, especially in nonregulated testing programs.

Diuretics may have the same effect as hydration or dilution. One common belief is that herbal teas ("Golden Seal" is a common example) will effectively protect an individual from detection. There is some historical truth in this belief. In the days when thin-layer chromatography was used to screen for drugs, some material in the tea acted as a masking agent by blocking the visualization of morphine or other opiates. The belief is reinforced by the fact that pigments in the tea result in a change in urine color to a darker hue. With the current immunoassay techniques, there is no value to the teas except the related diuresis and the increased hydration that results.

9.6 THE SCIENCE UNDERLYING THE SUBSTITUTED SPECIMEN

According to federal guidelines, a specimen is:

> Substituted (i.e., the specimen does not exhibit the clinical signs or characteristics associated with normal human urine) if the creatinine concentration is <2 mg/dl *and* the specific gravity is <1.0010 or >1.0200.

Specimens reported as "substituted" have creatinine and specific gravity values that are so diminished or incongruent that they are not consistent with normal human urine. Since the establishment of the federal guidelines for "substituted" specimen criteria, there have been an increasing number of specimens that have been reported as substituted — even some that had apparently been collected under direct observation. Many of these met the "substituted" criteria by having creatinine concentrations and specific gravity levels that were outside the acceptable range. More specifically these are "water-loaded" or "ultra-dilute" specimens.

What is the clinical basis for concluding that a urine specimen meets the current definition of "substituted"? Such a result is the administrative equivalent of an overt "refusal" to be tested or a "positive" drug test.

9.6.1 NLCP: State of the Science — Update 1

In February 2000, a technical review was published under the auspices of HHS/SAMHSA, entitled NLCP: "State of the Science – Update #1, Urine Specimen Validity Testing, Evaluation of the Scientific Data Used to Define a Urine Specimen as Substituted." Its purpose was to provide an overview of the published scientific literature that HHS used to develop the criteria for defining a specimen as substituted. The report's conclusion was as follows:

> In order for a specimen to be reported as substituted, both creatinine and specific gravity must meet defined criteria; that is, urine creatinine < 5 mg/dl *and* urine specific gravity < 1.001 or > 1.020. This testing requirement provides both an analytical and physiological safeguard. The review of the scientific literature including random clinical studies, medical conditions resulting in severe overhydration or polyuria, and water-loading studies confirms that the urine criteria of creatinine < 5 mg/dl *and* urine specific gravity < 1.001 or > 1.020 represent a specimen condition that is *not* consistent with normal human urine. In the deductive evaluation of 47 studies, no exception to the criteria defining a "substituted" specimen was reported.

However, the review cannot be interpreted as conclusive evidence that it is not possible for anyone to physiologically produce a "substituted" urine, since none of the studies cited was designed to look specifically at the amount of water it takes to produce a substituted specimen. Theoretical physiology suggests a lower value and nephrologists have more recently advised that on a mathematical basis individuals with low normal serum creatinine values may produce a urine creatinine value slightly below 5 mg/dl.

Later that same year, DOT performed a water-loading study that asked the question: How much water does it take to produce a substituted specimen? Or: How much water can a person consume and still not have a "substituted" specimen?

9.6.2   DOT Water-Loading Study

The DOT study,[7] entitled "Paired Measurements of Creatinine and Specific Gravity after Water Loading," is more on point. The study used 40 female and 13 male volunteers. The first phase was to simulate the shy bladder procedure. The first morning void was collected; then all volunteers were given 40 oz of fluid and were asked to provide a specimen every hour for the first 3 h and to continue with an additional 40 oz or more over the next 3 h. The volunteers were asked to drink as much as they could. Two participants were unable to consume the minimum amount of fluid intended. On the other hand, 11 participants (5 men and 6 women) consumed more than 1 gal of fluid by the end of their test periods. The bottom line was that none of the 480 specimens were identified as "substituted." The maximal suppression of creatinine values (which is the critical function value) was seen in an individual who had consumed 1.5 gal of water by the 5-h mark. It was reported that the creatinine value approached but was not below 5 mg/dl. Then, interestingly enough, the individual consumed another liter, and in the final hour had an elevated creatinine. Overall, the data showed that many of the volunteers attained levels of specific gravity below the 1.001 threshold value, but did not attain creatinine levels below 5 mg/dl. It, therefore, appeared that individuals cannot consume enough water to have a "substituted" specimen although it may be theoretically possible to do so. The results suggest that it is not physiologically possible to "unintentionally" drink too much water and be confronted with a "refusal to test–substituted specimen." In the absence of a related medical condition, unintentional ingestion of water seems to be as unlikely as passive inhalation of drugs. (Can a person unintentionally drink 1 1/2 gal of water?)

9.6.3   FAA Workplace Urine Specimen Validity Testing Colloquium

In light of several witnessed and documented cases of individuals demonstrating a creatinine concentration slightly below 5 mg/dl, the FAA sponsored a "Workplace Urine Specimen Validity Testing Colloquium," February 4–6, 2003. The colloquium was organized in direct response to a Congressional mandate to the FAA to prepare a report on whether there were any particular medical conditions, dietary factors, or individual characteristics that could result in a urine specimen meeting the existing criteria of an adulterated or substituted specimen. Participants included, among others, distinguished toxicologists and specific technical experts from the relevant fields of medicine, science, drug testing, and law. The participants had experience in drug-testing programs, medical, and other related fields. Presentations and discussions centered on the physiological impact of medical issues, working conditions, and dietary habits on validity testing conducted on specimens submitted under the DOT workplace program.

In April 2004, FAA released its report with the following findings and conclusions:

1. Dietary habits, medical issues, and working conditions do not affect the validity of specimens.
2. The validity testing criteria as currently established for adulteration are appropriate.

Only a few substituted specimens are attributed to physiological concentrations of urine creatinine (a compound in urine) below the HHS established value. However, the established value for creatinine is not appropriate for all people.

It made the following recommendations:

1. Substitution: The creatinine level for determining substituted specimens should be lowered to less than 2 mg/dl. This will prevent any individual who can achieve this concentration through normal physiological means from being improperly labeled as providing a substituted urine specimen.
2. Dilution: The current creatinine and specific gravity levels for determining a dilute specimen are adequate. For federally regulated programs, specimens identified as dilute should be tested for drugs at lower cutoff values in order to overcome the efforts of individuals to hide the presence of drugs.
3. Adulteration: The levels established for pH (<3 or >11), chromium VI (cutoff concentration of 20 µg/ml or more), and nitrite (greater than or equal to 500 µg/ml) are satisfactory. However, all laboratory testing for adulterants should include two distinct chemical methods to determine the presence of adulterants in the specimens (i.e., one for initial testing and another to confirm the presence of the adulterant).
4. Role of the Medical Review Officer: Supplement current verification process for assessing individuals' claims that they can produce a specimen meeting the substituted criteria through physiological means. Provide additional guidelines and training for the changes in the verification process.

9.7 ADULTERATION

"Adulterated" is the term used for a specimen that has been altered by the donor in an attempt to defeat the drug test. The donor's goal in this regard is to affect the ability of the laboratory to properly test the specimen for drugs and/or to destroy any drug or drug metabolite that may be present in the specimen.[8-10] Many substances can be used to adulterate a urine specimen *in vitro*, including common household products, commercial chemicals, and commercial products developed specifically for drug test specimen adulteration. Adulterants are therefore readily available, may be easily concealed by the donor during the collection procedure, and can be added to a urine specimen without affecting the temperature or physical appearance of the specimen. To identify adulterated specimens, HHS requires certified laboratories to perform a pH test and a test for one or more oxidizing compounds on all regulated specimens. Laboratories are also allowed to test regulated specimens for any other adulterant, providing they use initial and confirmatory tests that meet the validation and quality control requirements specified by the HHS Guidelines.

An adulterant may interfere with a particular test method or analyte, but not affect others. For example, an adulterant may cause a false-negative marijuana (cannabinoids) result using a particular immunoassay reagent, but not affect the test results for other drugs. The same adulterant may not affect the test results obtained using a different immunoassay reagent or method. It is also possible for an adulterant to cause a false-positive drug test result, rather than the intended false negative. The initial drug test required for federal workplace programs (immunoassay) is more sensitive to adulterants than the required confirmatory drug test (gas chromatography/mass spectrometry, GC/MS). Currently, the GC/MS assays for marijuana metabolite (THCA) and opiates appear to be the most affected.

An adulterated specimen is defined in 49CFRPart 40.3 as "A specimen that contains a substance that is not expected to be present in human urine, or contains a substance expected to be present but is at a concentration so high that it is not consistent with human urine." An adulterated specimen may be reported by the laboratory as having a nitrite concentration that is too high, a pH level that is too high or too low, or identified as having a specific adulterant (such as glutaraldehyde) present.

The reference in the HHS guidelines to a general oxidant test is new. Most adulterants used are oxidizing agents and the early tests for nitrite and chromate were oxidant tests, although specifically named as either nitrite or chromate. Since the general characteristic of the adulterants was that of an oxidant, it is practical to use a comprehensive general test for screening purposes to be followed by specific confirmation tests, rather than do a series of similar and potentially cross-reacting tests to screen for adulterants.

HHS allows certified laboratories to test for any adulterant. It is not possible to provide specific program guidance for all substances that may be used as adulterants; however, HHS has included specific requirements in the Guidelines for pH analysis and for the analysis of some known adulterants that follow.

9.7.1 pH

The pH of human urine is usually near neutral (pH 7) although some biomedical conditions affect urine pH. HHS set the program cutoffs for pH based on a physiological range of approximately 4.5 to 9. Specimens with pH results outside this range are reported as invalid. An extremely low pH (i.e., less than 3) or an extremely high pH (i.e., at or above 11) is evidence of an adulterated specimen. Urine specimens that are found to have a pH <3 or >11 can be reported as "Specimen Adulterated: pH is too high (or too low)." The establishment and publication of these threshold limits end the debate and uncertainty that has surrounded the issue of what is a normal urine pH and what margin of safety should be given to upper and lower levels.

To the average donor who may be contemplating the addition of a chemical substance to his to her urine, acids and bases seem like good choices. Battery acid and Drano have been used and are generally available. Such compounds have a significant impact on the pH of the specimen and include hydrochloric (muriatic) acid used for swimming pools. Commercial adulterants that affect pH include Amber-13 and THC-free. Amber-13 is a sulfur-smelling liquid sold in a glass vial. When 8 ml of Amber-13 is mixed with 50 to 150 ml of water it produces a pH of about 1. Urine will probably buffer this to some degree. THC-Free lists its formulation as water, muriatic acid, potassium chloride, phosphoric acid, and potassium hydrogen phosphate. One vial of THC-Free added to 50 m of urine will produce a pH of between 1 and 2.

9.7.2 Nitrite

Nitrite is an oxidizing agent that has been identified in various commercial adulterant products. Nitrite is produced by reduction of nitrate. Nitrite in high concentrations is toxic to humans, especially infants, causing methemoglobinemia by oxidizing the iron in hemoglobin. Nitrate and to a lesser extent nitrite are present in the environment. Nitrite may be normally present in human urine.

There has been a lot of confusion between nitrites and nitrates. It is important to understand the difference between the two compounds.

Nitrite is NO_2 — Nitrites have a therapeutic use as a vasodilator (oral dose = 30 to 60 mg). Inorganic nitrite can be used in treatment of cyanide poisoning (intravenous dose = 300 to 500 mg). One example of an organic nitrite is nitroglycerin. Nitrite is toxic; after accidental ingestion of nitrite, toxic symptoms include weakness, nausea, numbness, shortness of breath, tachycardia, and cyanosis. One reported case of nitrite overdose resulted in a urine concentration of 340 mg/l.

Nitrate is NO_3 — Nitrates have a widespread use as fertilizers, which can lead to accumulation in food plants and water supplies.

Both nitrites and nitrates are also used as curing agents in processed meats. They may legally be present in these products in the following concentrations: nitrites up to 200 ppm and nitrates up to 500 ppm. Some pathological conditions (infection, inflammation), medical treatments (cancer), and urinary tract infections may result in nitrites in urine.

In whole blood, nitrite is unstable. Whole-blood nitrite is almost completely converted to nitrate within 1 h. The nitrate that is formed is almost completely excreted in the urine. Normal nitrite concentrations in urine are very low and have been demonstrated in a variety of studies to generally be below 100 to 200 µg/ml.

Because low levels of nitrite may be present in human urine, HHS set a cutoff level of 500 µg/ml for adulteration and 200 µg/ml for invalid results. These concentrations are well above levels normally seen in human urine. Therefore, normal exposure does not explain a nitrite-adulterated result.

Drug testing laboratories began to study the effects of nitrite in detail when the product Klear came on the market in 1997. A urine specimen would screen positive and then fail confirmation. It would not just fail in the sense that no drug was identified, but it looked like something was destroying all of the organic compounds in the specimen.

Specimens adulterated with Klear were oxidizing THCA during the extraction process. When aliquots to be tested were acidified, the oxidation process exponentially accelerated. The mass ion peaks on the mass spectral chromatograms were absent. There is no drug and there is no internal standard either. The internal standard is a deuterated version of the drug or metabolite being tested. It is added as a control reference for the analysis and its destruction is diagnostic for an oxidizing adulterant.

Some commercial nitrite adulterants include Klear, Whizzies, and Randy's Klear (I). Klear is a product that comes in a small plastic vial containing 500 to 800 mg of white crystals (potassium nitrite). The cost of two tubes is $29.95. Whizzies (sodium nitrite) appears to be a knockoff product of Klear. It is sold as white powder contained in two vials each containing about 850 mg of the compound.

9.7.3 Chromium

Chromium exists in a number of chemical states. The zero valence state, Cr^0, is the metallic state. Chromium also exists in nature in a divalent [Cr^{2+}, Cr(II)], trivalent [Cr^{3+}, Cr(III)], and hexavalent [Cr^{6+}, Cr(VI)] state. Understanding the significant differences in the chemistry and toxicology is important in understanding and interpreting adulteration results. The divalent state is very unstable and is rapidly changed to Cr^{3+}. The biologically or toxicologically significant states are the trivalent Cr^{3+} and hexavalent Cr^{6+} states.

Trivalent chromium Cr^{3+} is an essential nutrient in diet. It plays a critical role in maintaining normal glucose tolerance. The trivalent chromium is the species found in the dietary supplement chromium picolinate. It is important to realize that even large doses of chromium picolinate will not produce any hexavalent chromium in the urine. It will, naturally, produce a small amount of trivalent chromium in the urine. A literature review shows levels in the low nanogram range. The highest concentration of trivalent chromium reported was 11 ng/ml. Those individuals took 400 µg/day of chromium picolinate for 3 days.

The hexavalent chromium Cr^{6+} is a strong oxidizing agent. Cr^{6+} is reduced to Cr^{3+}; however, Cr^{3+} is not converted to Cr^{6+}. Cr^{6+} is used in chrome plating, dyes and pigments, leather tanning, and wood preserving. Hexavalent chromium is present in the environment and is a carcinogen. It is a very toxic and irritating compound. In studies that looked at dietary or environmental exposure to hexavalent chromium, the maximum level reported was 690 ng/ml. In this instance, 10 mg/day of Cr^{6+} (in water) was ingested for 3 days. These levels are in stark contrast to what is seen with adulterated urine.

In summary, concentrations of Cr^{3+} in urine are small — in the nanogram-per-milliliter range. Concentrations of Cr^{6+} found in adulterated urine specimens are large — in the microgram-per-milliliter range. In addition, the laboratory adulteration assays are specific for Cr^{6+} and do not include Cr^{3+}. Urine concentrations of hexavalent chromium found in adulterated urine specimens exceed the highest toxic case reported. (The case was suicide by ingestion of chromic acid solution; 3 days after ingestion, urine contained 5.13 µg/ml. Death occurred 1 month later from injury.) The HHS reporting cutoff is 50 µg/ml.

Some chromium containing adulterants include Klear II, LL418, Sweet Pee's Spoiler (pyridium chlorochromate), UrineLuck, and Ultra Kleen (chromate).

Pyridinium chlorochromate is a strong oxidizing agent that has been the agent in some commercial adulterants. This compound is identified by urine drug testing laboratories using a confirmatory test for pyridine. Pyridine at any detectable level in a urine specimen is evidence of adulteration.

9.7.4 Hydrogen Peroxide and Peroxidase

The commercial adulterant Stealth comes in a packet that contains two plastic vials. One vial contains a tan solid that does not melt. The other vial contains about 1.7 ml of a clear liquid, with a measured pH of 4.5. The user is instructed to pour the solid into the collection cup, add urine, then add the liquid "activator." The tan solid contains a peroxidase, the enzyme that speeds the oxidation process. The liquid contains peroxide. Stealth goes to work quickly and oxidizes the THC metabolites in a matter of hours, then "self-destructs." Hence, the name Stealth.

9.7.5 Halogens

Halogens are the four elements fluorine, chlorine, bromine, and iodine. Halogen compounds have been used as adulterants. The term "halogen" (from the Greek *hals*, "salt," and *gennan*, "to form or generate") was given to these elements because they are salt formers. None of the halogens can be found in nature in its elemental form. They are found as salts of the halide ions (F^-, Cl^-, Br^-, and I^-). Fluoride ions are found in minerals. Chloride ions are found in rock salt ($NaCl$), the oceans, and in lakes that have a high salt content. Both bromide and iodide ions are found at low concentrations in the oceans, as well as in brine wells. The assays used by certified laboratories identify halogen compounds that act as oxidants. These do not include the halogen salts that may be present in a urine specimen. The presence of an oxidative halogen in a urine specimen is evidence of adulteration.

Iodine/iodate is a recent adulterant in the halogen class. Molecular iodine (I_2) is a blue-black solid that sublimes. Iodide (I^-) is generally found in the form of potassium or sodium salts (KI, NaI). Iodate (IO_3^-) also appears as potassium or sodium salts (KIO_3, $NaIO_3$) of iodic acid. Iodate is a strong oxidizing agent and is reduced to iodide. Iodide and iodate are used as food and salt additives, in antiradiation products and in thyroid hormones. Normal concentrations are approximately 600 µg/L.

9.7.6 Glutaraldehyde

Glutaraldehyde is a clear, colorless liquid with a distinctive pungent odor sometimes compared to rotten apples. One of the first effective commercial adulterants, UrinAid, was found to contain glutaraldehyde. Glutaraldehyde is used as a sterilizing agent and disinfectant, leather tanning agent, tissue fixative, embalming fluid, resin or dye intermediate, and cross-linking agent. It is also used in X-ray film processing, in the preparation of dental materials, and surgical grafts. Glutaraldehyde reacts quickly with body tissues and is rapidly excreted. The most common effect of overexposure to glutaraldehyde is irritation of the eyes, nose, throat, and skin. It can also cause asthma and allergic reactions of the skin.

Glutaraldehyde does not normally occur in urine and is readily detectable. It will interfere with the immunochemical screening tests. It may also interfere with the recovery of the drug in GC/MS analysis or it may destroy the metabolite. It is interesting to note that, although this product is sold as a way to guarantee a negative in the urine of marijuana users, it affects the analysis of other drugs as well. It has a chemical aldehyde smell and it will denature proteins in the urine, so that over a 1- or 2-day period of time a brown precipitate will appear. No matter what the preliminary basis may be for suspecting adulteration of a specimen by glutaraldehyde, the presence of glutaraldehyde should be confirmed by reliable chemical analysis such as GC/MS. Glutaraldehyde at any detectable level in a urine specimen is evidence of adulteration.

Clear Choice is another adulterant containing glutaraldehyde and squalene.

9.7.7 Other Chemicals and Household Products

Surfactants, including ordinary detergents, have been used to adulterate urine specimens. Surfactants have a particular molecular structure made up of a hydrophilic and a hydrophobic component. They greatly reduce the surface tension of water when used in very low concentrations. Foaming agents, emulsifiers, and dispersants are surfactants that suspend an immiscible liquid or a solid, respectively, in water or some other liquid.

Liquid Soap: Reported to cause false negatives in EIA procedures for THC and PCP. Its addition increases the pH of the specimen. A commercial adulterant called Mary Jane SuperClean 13 was reported to be primarily soap.

Sodium Chloride: Reported to cause false negative in EIA procedures. Its addition increases the specific gravity and chloride ion content of the specimen.

Bleach: Reported to cause a false negative in EIA procedures for all five drugs tested for by SAMHSA and in FPIA procedures for all but cocaine.

Drano: Reported to cause false negatives in EIA procedures. Its addition increases the pH of the specimen.

Sodium Bicarbonate: Reported to cause false negatives in EIA procedures for opiates and PCP. Its addition increases the pH of the specimen.

9.8 THE INVALID RESULT

HHS describes an invalid result as follows:

When a laboratory is unable to obtain a valid drug test result or when drug or specimen validity tests indicate a possible unidentified adulterant, the laboratory reports the specimen to the MRO as "invalid result."

This definition is not a comprehensive. Invalid results also include a category of specimens that can be described as having suspect or abnormal characteristics, such as pH out of the normal range, high nitrite levels, or unusual creatinine and specific gravity levels. It is true that "invalid" results can indicate a possible unidentified adulterant or substitution and that many invalid results could not be physiologically possible. But in many cases there is just an absence of conclusive scientific evidence that the specimen is not physiologically possible, and/or that the analytical method is not definitive, precise, or valid enough to withstand legal challenge.

When an MRO receives an "invalid" specimen report, it is incumbent upon him or her to discuss with the laboratory whether additional tests should be performed by the laboratory or by another certified laboratory. It may be possible to obtain definitive drug test results for the specimen using a different drug test method or to confirm adulteration using additional specimen validity tests. The choice of the second laboratory or additional tests will be dependent on the suspect adulterant and the validated characteristics of the different drug test. Laboratory staff should be knowledgeable of their tests' validated characteristics including effects of known interfering substances, and be able to recommend whether additional testing is worthwhile.

The current HHS specimen validity rules require screening of specimens for oxidants, but do not require the laboratory to confirm the screening results. The rationale is that requiring specific confirmation methods for a broad class of defined and undefined adulterants would represent a significant cost increase in laboratory services. The majority of laboratories have decided not to bother with confirming oxidant results, and simply report the screening results as "invalid." Somewhat unanticipated is that even laboratories that were confirming common oxidants such as nitrites have withdrawn from this practice. The specimen validity rule requires the laboratories to have more comprehensive and definitive confirmatory procedures for nitrites. This likely requires the acquisition of additional laboratory equipment. In addition, there is significant time involved in confirming results of any type. In the wake of increased litigation, there are lingering liability concerns at the laboratories. The end result is that only a small number of laboratories perform confirmatory procedures for nitrites and other oxidants as required by the current rules.

So today there have evolved three conceptual categories of laboratory results: the positive (drug, adulterated, and/or substituted), the invalid (everything in-between), and the negative (dilute and non-dilute). Each requires the focus and attention of the laboratory and the MRO.

REFERENCES

1. Cook, J.D., Caplan, Y.H., LoDico, C.P. and Bush, D.M. The characterization of human urine for specimen validity determination in workplace drug testing: a review. *J. Anal. Toxicol.*, 24: 579–588 (2000).
2. U.S. Department of Transportation, Office of the Secretary, Procedures for Transportation Workplace Drug and Alcohol Testing Programs, 49 CFR Part 40, Final Rule (65 FR 79462), December 19, 2000.
3. U.S. Department of Transportation, Office of the Secretary, Procedures for Transportation Workplace Drug and Alcohol Testing Programs, 49 CFR Part 40, Technical Amendment, Final Rule (66 FR 41944), August 9, 2001.
4. U.S. Department of Health and Human Services, Substance Abuse and Mental Health Services Administration, Mandatory Guidelines for Federal Workplace Drug Testing Programs (69 FR 19644), April 13, 2004.
5. U.S. Department of Transportation, Office of the Secretary, Procedures for Transportation Workplace Drug and Alcohol Testing Programs, 49 CFR Part 40, Interim Final Rule (68 FR 31624), May 28, 2003.
6. U.S. Department of Transportation, Office of the Secretary, Procedures for Transportation Workplace Drug and Alcohol Testing Programs, 49 CFR Part 40, Interim Final Rule (69 FR 64865), November 9, 2004.
7. Edgell, K., Caplan, Y.H., Glass, L.R. and Cook, J.D. The defined HHS/DOT substituted urine criteria validated through a controlled hydration study. *J. Anal. Toxicol.* 26: 1–5 (2002).
8. Shults, T.F. The MRO's role in managing adulteration, substitution, dilution, refusal and other problems. In *The Medical Review Officer Handbook*, Quadrangle Research, Research Triangle Park, NC, 1995, 129–159.
9. Caplan, Y.H., Urine Specimen Validity. Medical Review Officers Training Program, American Association of Medical Review Officers, Chicago, Illinois, June 25, 2005.
10. Shults, T.F., Update on adulterants. *MRO Alert.* 16(1): 6–9 (Dec. 2004/Jan. 2005).

CHAPTER 10

The Role of the Medical Review Officer in Workplace Drug Testing

Joseph A. Thomasino, M.D., M.S., FACPM
JAT MRO, Inc., Jacksonville, Florida

CONTENTS

10.1 THE MRO AS THE "GATEKEEPER" OF THE WORKPLACE DRUG TESTING PROGRAM

A Medical Review Officer (MRO) has come to be defined in U.S. Department of Transportation (DOT) regulations (i.e., 49 CFR Part 40) as a licensed physician (Doctor of Medicine or Osteopathy) who is knowledgeable about and has clinical experience in controlled substance abuse disorders, including detailed knowledge of alternative medical explanations for laboratory confirmed drug test results. The MRO has become an integral part of the workplace drug testing process as federal regulations for workplace drug testing have been developed and implemented, at first for drug testing of federal employees, and then for millions of other workers in private industry for which drug testing was mandated by federal agencies such as the U.S. DOT, The U.S. Coast Guard, and the Nuclear Regulatory Commission. Beginning in the mid-1980s the MRO has been involved in an ever-growing number of drug tests. Federal regulatory requirements for workplace drug testing have expanded and these regulations have further defined and broadened the role of the MRO in the process. Programs for workplace drug testing requiring medical review have been implemented by various states (e.g., Florida, Georgia, others) in connection with worker compensation programs. Some states have required medical review of all workplace drug test results collected in those states (e.g., Oklahoma and New York). Increasing numbers of private employers have been implementing drug testing programs that include medical review of results absent any regulatory requirement to do so. Although originally all workplace drug testing programs involved the collection of urine drug testing specimens, and most still do, other sampling media including hair, blood, saliva, and sweat, and others are beginning to be accepted and used in workplace drug testing programs. All of these factors have led to the emergence of the MRO as the "gatekeeper" of the workplace drug testing process.

The identification of the MRO as the "gatekeeper" in the workplace drug testing process was first made in DOT regulations. It is an apt characterization of the overall role of the MRO in this process. One way to conceptualize this role is to consider that the MRO is an "equity agent." In a sense the MRO oversees all the elements of the process. The MRO ensures that the donor (i.e., the individual providing the drug testing specimen) has had the specimen collected properly, that it has been analyzed correctly, that the result has been reported to the MRO promptly and clearly, that alternative medical explanations for any positive or other non-negative results where appropriate have been sought, that the donor's technical questions have been answered and the donor's response to an adverse determination (e.g., a reconfirmation test, substance abuse professional evaluation, etc.) has been facilitated, and that a prompt and clear report of the final determination/verification has been made to the employer or other organization commissioning the drug test. By performing the above functions in reviewing each drug test the MRO helps confer legitimacy and fairness to a process that can become contentious as serious sanctions are often applied to donors with verified non-negative results. A verified positive result, or a determination that the donor has refused to provide an adequate specimen for drug testing, may result in denial or termination of employment, loss of other benefits, or other adverse outcomes for the donor. Thus the role of the MRO in allowing the donor due process to explain properly collected and technically valid drug testing results before an adverse determination is made has become vital in maintaining the integrity, value, and effectiveness of workplace drug testing programs.

10.2 THE MRO AND THE COLLECTION PROCESS

Proper collection of drug testing specimens is the first and arguably the most critical step in the implementation of any workplace drug testing program. Important in this is that chain-of-custody procedures be established and followed in collecting drug testing specimens. Given the serious consequences for the donor that often follow adverse determinations, and the concomitant risk for the organization commissioning the test in applying those consequences inappropriately, it is clear that before becoming involved in the review of any drug test the MRO must be sure that appropriate chain of custody procedures were followed in the collection and further handling of the specimen as it is transported to and analyzed at the laboratory. Strict adherence to the appropriate collection protocol with close attention to properly establishing, maintaining, and documenting the chain of custody for each specimen is therefore vital to the overall success of any drug testing program.

The MRO is rarely, if ever, actually present at the time any specimen is collected. In most instances the MRO does not actually know, and has never met the collector for any given specimen. Therefore as the "gatekeeper" of the process the MRO must rely on review of the chain of custody and control form (CCF) that has been completed by the collector and the donor at the time the specimen is collected. Needless to say, prompt provision of a legible copy of the CCF to the MRO by the collector is absolutely vital for the proper management and promulgation of any drug testing program. Practically speaking, assuring that a legible copy of the CCF for each specimen is obtained as rapidly as possible after a specimen has been collected is a major activity for the MRO and staff. The success of most MRO practices, and the ability to gain and maintain clients by the MRO, are largely determined by how rapidly the MRO is able to provide reports of final determination/verification to the organization commissioning a test, once a test has been collected.

Once received, the CCF for each specimen is reviewed by the MRO or staff to ensure that it is legible and that it has been appropriately completed by the collector and the donor. The legible, properly completed CCF is then matched with the result received from the laboratory and the MRO and/or staff complete the review and make report to the organization commissioning the test. Any deficiencies in the CCF should be corrected, if possible, by the MRO before the final determination is made. Often deficiencies will have been detected and corrected by the laboratory before the result is reported to the MRO. However, if it becomes apparent that the laboratory has failed to detect or correct a deficiency, it falls to the MRO to do so.

Certain flaws on the CCF are "fatal." The most common of these are quantity of specimen insufficient for testing; no collector printed name or collector signature in the collector certification portion of the CCF; tamper-evident seal broken or missing on the specimen container upon arrival at the laboratory; and donor identification on the specimen container not matching the donor identification on the CCF submitted with the specimen. When these "fatal" deficiencies or flaws are detected the laboratory will not perform the analysis. When this condition is reported to the MRO the final determination/verification must be that the test was canceled due to the flaw. However, for other deficiencies, if the laboratory has not corrected them, the MRO must do so. These include, among others, a CCF on which the collector's name is printed but the collector has failed to sign that section of CCF in which the collector certifies that the specimen was submitted by the donor in question and that the specimen was collected, labeled, and sealed in accordance with the appropriate protocol; the certifying scientist has failed to sign the laboratory copy of the CCF when reporting the result (this becomes apparent to the MRO only in the case of a non-negative result for which the laboratory copy of the CCF must be provided to and reviewed by the MRO prior to making a final determination/verification); and times and/or dates on the CCF missing or contradictory. There is at least one flaw or deficiency that cannot ordinarily be corrected by the laboratory prior to analysis. This is failure of the donor to sign that section of the CCF in which the donor certifies that the specimen was actually submitted by the donor, that it has not been adulterated or tampered with, that the specimen container was sealed with a tamper-evident seal in the donor's presence, and that the information on the CCF and the label affixed to the specimen container is correct. As the laboratory ordinarily does not receive a copy of the CCF that bears the donor's signature, it is the MRO's responsibility to check for and correct this deficiency when it arises. Unless the collector has noted in the remarks section of the CCF that the donor has refused to sign, or that the collector forgot to have the donor sign before leaving the collection site, an attempt should be made to correct the flaw. This flaw, like other flaws that can be corrected, is corrected by having the collector sign an affidavit, certificate of correction, or memorandum for record (these are synonymous terms). This document should indicate that despite the flaw, the specimen was, in fact, submitted by the donor in question, the specimen was otherwise collected, handled, and transported to the laboratory properly, and that this is a true and accurate statement on the part of the collector. Once the properly executed document is received by the MRO it is maintained with the other documentation for the specimen, and review proceeds as for any other specimen. If the deficiency is not corrected in a reasonable time, usually considered to be no more than 1 to 2 weeks from when the flaw was detected and the certificate of correction was presented to the collector, the specimen is reported as canceled due to the uncorrected flaw.

When "fatal" or uncorrected flaws result in the cancellation of a specimen, it is the responsibility of the MRO to document and point this out to the collector in question, urge the collector to ensure that the situation is not repeated, and suggest that if additional training or education is needed to address the situation and prevent recurrences that it be promptly obtained by the collector. Even minor deficiencies or administrative mistakes that do not cause cancellation of the test, or require formal correction per se in order to make a final determination/verification, should be documented and pointed out to the collector for corrective action. Errors of this sort that have no significant adverse effect on the donor's ability to have a fair and accurate drug test include, among others, failure of the collector to indicate whether the specimen temperature was read within 4 min of collection and/or whether the temperature was within range; minor mistakes in recording the donor identification number on the CCF; reason for test inappropriate or unmarked on the CCF; failure to directly observe a specimen in instances where observation was called for; and delay in the collection process.

By careful review of the CCF for each specimen the MRO helps ensure the integrity of the collection process, and by extension the entire drug testing process, as proper collection is the foundation of any successful workplace drug testing program. By correcting deficiencies that are detected and making collectors aware of them, the MRO strengthens and fosters this most important element of workplace drug testing programs.

10.3 THE MRO AND THE ANALYTICAL LABORATORY

In a sense the MRO is one of the main "customers" of the analytical laboratory performing toxicological testing in the drug testing process. Although the organization commissioning the test may actually originally arrange for and pay for drug testing, in most instances the analytical laboratory reports results directly to the MRO who makes a final verification of the results before reporting them to the organization commissioning the test that has engaged the MRO for this purpose. In most instances where an organization has engaged an MRO the organization is not made aware of the laboratory-confirmed result by the laboratory and the laboratory will not reveal the laboratory-confirmed result to the organization, insisting that the MRO report all results to the organization. This arrangement is required for federally mandated testing and by some state-sponsored testing programs, e.g., in Florida. In other states either all results or only laboratory-confirmed positive results (e.g., in Maryland) must be reviewed by the MRO before they are reported to the organization commissioning the test. Therefore, in practice, the MRO and staff usually develop close working relationships with their counterparts at the analytical laboratories, i.e., the toxicologists, certifying scientists, and client service representatives. This is necessary to ensure prompt receipt of results by the MRO once analysis has been completed and for the MRO and staff to clarify and fully understand all aspects of the laboratory report.

As the "gatekeeper" in the drug testing process the MRO has a responsibility to correct any flaws in the testing process that may be discovered on the part of the analytical laboratory before making a final determination for any given test result. Failure of the certifying scientist to provide necessary documentation (e.g., the properly completed laboratory copy of the CCF for non-negative results on federal testing), failure of the certifying scientist to sign or otherwise fully and properly complete necessary documentation, and correction of flaws in the collection process not detected by the analytical laboratory prior to release of results to the MRO are examples of this.

The MRO has a responsibility to ensure that all aspects of the laboratory report are understood before making a final verification. This is particularly important when invalid drug test results are reported by the laboratory. In these instances, where for a variety of technical reasons the toxicologist does not feel that a reliable analysis can be made (e.g., presence of an interfering substance the exact nature of which is unknown, urine specimen colored blue or having some other unusual color or appearance, urine creatinine low with normal specific gravity, among others), the MRO must interview the donor to establish whether a legitimate medical explanation can be established to explain these circumstances. It is incumbent on the MRO to fully understand why the specimen was deemed unsuitable for analysis rendering the invalid result, and to consult with appropriate laboratory personnel in this regard, so as to properly direct the interview with the donor. Similarly, there are instances where it is not clear to the MRO whether a donor's explanation for any non-negative result would in fact explain the result. The MRO can and should consult with appropriate laboratory personnel in these instances and make reliance on the information and guidance they provide in these matters.

Finally, the MRO often has a responsibility to offer the donor the option to have a drug testing specimen with a non-negative result sent to a different analytical laboratory for reconfirmation testing after an adverse determination has been made for that result by the MRO. This is required for federally mandated drug testing, by some state-sponsored drug testing programs, by state law in some states, and as a matter of organizational policy for some organizations commissioning drug testing. As the "gatekeeper" of this process, if a specimen fails to reconfirm for the result in question, the MRO has a responsibility to then amend the final determination/verification for that specimen. The MRO must also notify the donor and the organization commissioning the test of this, and depending on the type of testing, e.g., federally mandated, state sponsored, etc., may also have to make a report to a governmental or other body overseeing the drug testing process. Such a report may have serious consequences for the analytical laboratory found to have reported a "false-positive" or other erroneous result in terms of the laboratory's ability to continue to provide drug testing services.

In practice the MRO and staff must develop and maintain good communications and close working relationships with their counterparts at the analytical laboratories. This is essential for the proper conduct of any drug testing program.

10.4 THE MRO AND THE VERIFICATION OF DRUG TESTING RESULTS

The primary role of the MRO in workplace drug testing programs has always been to verify the results of the drug tests collected. When federally mandated drug testing programs were first implemented in the 1980s it was considered essential that donors with laboratory-confirmed positive drug testing results be afforded an opportunity to confidentially provide a legitimate alternative medical examination, if one existed, before the commissioning organization learned of the result. In the event that a legitimate alternative medical explanation could be established, the laboratory-confirmed positive result would be "downgraded," i.e., reported as negative to the commissioning organization by the MRO in such a way that it would appear to be no different from any other laboratory-confirmed negative result. As part of allowing "due process" before sanctions were applied to the donor as a result of a laboratory-confirmed positive result, the role of the MRO as the finder of fact, interpreter of information offered, and maker of the final determination was conceived.

A large portion of the time an MRO devotes to MRO practice is spent conducting and interpreting the results of interviews conducted with donors with laboratory-confirmed non-negative results. Originally, these were individuals with laboratory-confirmed positive results for the five drugs or classes of drugs tested for in federally mandated testing, i.e., amphetamine and methamphetamine, cocaine metabolites, marijuana metabolites, opiates (specifically codeine, morphine, and 6-acetylmorphine), and phencyclidine. As non-federally regulated drug testing programs were implemented, expanded, and in some cases tailored to the needs of individual organizations, other substances including alcohol, barbiturates, benzodiazepines, methadone, methaqualone, other opiates, propoxyphene, and others have been included in workplace drug testing programs, and naturally the MRO has had to deal with laboratory-confirmed positive results for these as well. In this process the MRO confidentially conducts and documents the interview; explains the result to the donor and the MRO's role in the process; allows the donor the opportunity to present an explanation; gives the donor reasonable time to develop and provide any evidence supporting any explanation offered; and promptly reviews, interprets, confirms, and verifies any information received before making the final determination/verification.

Besides laboratory-confirmed positive results, other non-negative results have also come to require an interview of the donor and interpretation of information gathered prior to verification by the MRO. For federally mandated testing, urine specimens that have been adulterated with a foreign substance that can be specifically and reliably identified are reported as such to the MRO by the laboratory. The MRO conducts an interview to determine if a legitimate alternative medical explanation can be established for the presence of the adulterant in the specimen. If this cannot be established, the result is reported as a "refusal to test" to the organization commissioning the test with resultant sanctions applied to the donor as prescribed in federal regulations. For federally mandated testing, urine specimens with extremely low creatinine levels (i.e., 2 mg/dl or less) and specific gravity of 1.001 or less or 1.020 or greater are reported as substituted (i.e., not consistent with normal human urine) to the MRO. Similar to adulterated specimens, for substituted specimens the MRO must also conduct an interview to determine if a legitimate alternative medical explanation can be established to explain these abnormal creatinine and specific gravity values. If it appears to the MRO that such an explanation may exist, the donor is then required to demonstrate under observed and controlled conditions that urine with these abnormal characteristics can once again be produced. The MRO reviews the results of this procedure in making the final determination. Once again, failure to establish an acceptable explanation or to demonstrate the production of urine meeting these criteria under observed and controlled circumstances will result in a final verification of the result and report to the commissioning organization as a "refusal to test." Invalid drug test results, as discussed above, also require an interview with the donor. In these if a legitimate alternative medical explanation can be established, the result is simply verified as canceled with no further action required unless a negative result is required (on federally mandated testing this is for pre-employment, return to duty, or follow-up testing). If a legitimate alternative medical explanation cannot be established, the test is verified as canceled but on federally mandated testing the organization commissioning the test is informed that an immediate re-collection directly observed by an individual of the same sexual gender as the donor must be conducted with minimal advance notice to the donor.

There are some additional circumstances that require review by the MRO of information not gathered directly from the donor or provided by the analytical laboratory. Results of evaluations conducted in response to substituted specimens were mentioned above. For opiate-positive results (i.e., codeine or morphine) the MRO in some circumstances (6-acetylmorphine negative, and codeine or morphine level less than 15,000 ng/ml on federally mandated urine drug testing) must have clinical evidence of opioid abuse before making a final determination/verification that the test is positive. This involves in some cases a "hands on" physical examination conducted by the MRO, or review by the MRO of an examination conducted by another physician acceptable to the MRO, to establish whether there is clinical evidence of opioid abuse (e.g., needle marks or tracks, disturbance of the sensorium, neurological abnormalities, etc.) before making the final determination/verification. Individuals who do not provide sufficient urine for testing even when provided extra time and fluids to do so are putatively demonstrating so-called "shy bladder." In these cases, on federally mandated testing, the donor is required to undergo an examination by the MRO or other qualified physician acceptable to the MRO to establish whether an ascertainable physiologic condition (e.g., a urinary system dysfunction) or a documented preexisting psychological disorder had or with a high degree of probability could have prevented the donor from providing a sufficient quantity of urine for testing. The MRO must review the results of this evaluation, seriously consider the opinion of the examiner if the examination was conducted by another, and render a final determination. If there is an acceptable explanation established as a result of the examination the test is verified as canceled unless a negative result is required; if there is no acceptable explanation established it is reported as a "refusal to test." On federally mandated testing where a negative result is required and the test is canceled due to an established legitimate explanation, an additional examination of the donor is required to establish whether clinical evidence of illicit drug use also exists. The results of this examination, conducted once again either by the MRO or another physician acceptable to the MRO, are reviewed by the MRO and a final determination is rendered. This evaluation may also include drug testing using another medium such as blood, saliva, hair, etc. If there is no evidence of illicit drug use, the test is verified as negative and is so reported to the organization commissioning the test. If there is evidence of illicit drug use, the test is reported as canceled and the evidence of illicit drug use is also reported to the organization commissioning the test.

Although the personal focus of the MRO is naturally on non-negative results, the vast majority of results reviewed by the MRO and staff are negative. The MRO has a responsibility to ensure that negative results are also properly reviewed and promptly verified and reported as negative. The concerns expressed above concerning collection and laboratory issues apply to negative results as well. Regular, periodic review by the MRO of negative results verified by staff is required for federally mandated testing, and is an essential element of good MRO practice in general.

The heart of the MRO's practice and professional responsibility is timely, accurate, and equitable review, verification, and reporting of drug testing results. The integrity, credibility, and success of this vital public safety and health program depend on the responsible and proper performance of this function by professionals dedicated to it.

10.5 THE MRO AND SAFETY ISSUES

Often, during the interviews conducted for non-negative results, the MRO will be informed by the donor, or will otherwise learn through other information provided as part of this process, of a condition the donor is suffering or a medication the donor is taking, that either renders the donor medically unqualified for safety-sensitive work in terms of the federal regulation in response to which drug testing is being conducted or which otherwise poses a potential safety hazard to the worker, co-workers, or the general public. Many of these federal regulations, e.g., those of the DOT, require the MRO to report these circumstances to the employer or other organization commissioning the drug test, without the consent of the donor. Some state-sponsored drug testing programs, e.g., the Florida Drug-Free Workplace Program, include similar provisions. In general, it is common for the MRO to feel obligated to report such circumstances to the organization commissioning drug testing even when not required or encouraged to do so by any extant federal or state law, regulation, or guideline.

Obviously, such a report will usually have serious consequences for the donor as it may result in the donor being prohibited from performing all or part of the donor's work duties until the situation is resolved. DOT regulations, for example, currently require the MRO to report the safety issue to the employer. However, these regulations also require the MRO to instruct the donor that if additional information is provided to the MRO in a timely fashion from the donor's treating health practitioner who modifies the situation, the MRO must then share that with the employer for the employer's further consideration of the matter. A similar approach may be adopted by the MRO for non-federally mandated testing in an attempt to resolve these potentially serious and often contentious issues.

This issue has been complicated by the Health Insurance Portability and Accountability Act (HIPAA). Some information obtained during the MRO interview or other portions of the drug testing process may be considered "protected health information" and therefore under HIPAA would require specific consent or authorization from the donor before it was released by the MRO to the organization commissioning the test. The DOT has opined that such information gathered as part of the drug testing regulated by that agency is exempt from the provisions of HIPAA, and that the MRO, and other service providers, need no consent or authorization to release information gathered during this process in accordance with the federal regulation. However, for state-sponsored programs or other non-federally mandated testing, it is not at all clear that "protected health information" gathered during the process of drug testing can be released by the MRO or other service providers to the organization commissioning the test in the absence of the consent or authorization of the donor without running afoul of HIPAA. In some states, for example, Maryland, certain information gathered as part of the drug testing process cannot be released to the organization commissioning the test without the donor's consent or authorization as a matter of state law, over and above any requirements imposed by HIPAA. Until this issue is clarified, and in some states regardless of any interpretations or modifications of HIPAA, it would appear most prudent for the MRO to obtain specific written consent from the donor before releasing "protected health information" to the organization commissioning the test, unless as is the case for DOT testing the issue has been specifically and definitively addressed. In practice, this may be accomplished *ad hoc* on a case-by-case basis as the need arises, or by advising the organization commissioning the test to require an appropriate written consent or authorization be executed by the donor before drug testing is performed.

10.6 THE MRO AND OTHER ADMINISTRATIVE FUNCTIONS IN WORKPLACE DRUG TESTING

The MRO may be called upon to perform a number of administrative functions beyond simply reviewing and reporting of results.

As discussed above, additional examinations of donors may be required before the MRO can make a final determination/verification in certain instances. For federally mandated testing donors with shy bladders (i.e., not able to produce a sufficient quantity of urine for testing), donors whose inability to produce a sufficient quantity of urine for testing is due to a permanent or long-term condition and for whom a negative result is required (i.e., on pre-employment, follow-up, or return to duty testing), those with adulterated or substituted specimens under certain circumstances, and donors for whom clinical evidence of opioid abuse must be obtained, will require the MRO to approve of the selection of the referral physician other than the MRO for these examinations, and in some instances help either the organization commissioning the test or the donor to locate a suitable examiner.

On federally mandated testing the MRO has a responsibility to cooperate with the Substance Abuse Professional (SAP) working with a donor for whom the MRO has verified a result as positive, or refusal to test. The MRO must also provide available information that the SAP requests, e.g., quantitative test results, information gathered during the MRO interview, etc.

The requirement on federally mandated testing for the MRO to report medical information to the organization commissioning the test that is likely to result in the donor being determined to be medically unqualified for safety-sensitive duties, or otherwise indicates that continued performance of safety-sensitive duties by the donor is likely to pose a significant safety risk, has also been mentioned above. Prompt sharing of additional information received from the donor's physician that might modify such a situation with the organization commissioning the test is an important administrative function the MRO is called upon to perform.

For all forms of testing the MRO has a responsibility to preserve the confidentiality of all information gathered and maintain records of MRO activities for varying periods of time depending upon regulatory requirements, state laws, contractual obligations, and professional guidelines. The MRO may also be called upon to provide these records to individual donors and produce them in various court or other legal or administrative proceedings.

Finally, in practice, the MRO is often called upon as a general consultant source for all matters dealing with drug and alcohol testing, not only by employers, but also by labor organizations and individual donors with concerns about drug testing. The MRO may be called upon by the organization commissioning the test to help arrange and review specialized toxicological testing tailored to some particular circumstance or situation in the workplace. Perhaps the most important administrative function for the MRO is to be knowledgeable and available to employers and donors to address their questions and concerns so as to support the integrity and credibility of the drug testing process.

10.7 EMERGING ISSUES FOR THE MRO

A number of issues are currently emerging that will affect MRO practice. These include specimen validity issues, alternative testing matrices, expanding the scope of toxicological testing on federally mandated and other testing, and on-site testing.

Concerns have been raised as to the level of creatinine in the urine that when present with a specific gravity of 1.001 or less or 1.020 or greater is to be considered evidence of a substituted specimen, not consistent with normal human urine, and thus exposing the donor to sanctions if such a specimen is verified as a refusal to test on this basis. Formerly this level was 5 mg/dl or less. However, quite recently, in response to scientific review of this issue and demonstrated ability on the part of donors to produce creatinine levels of 3 to 4 mg/dl simply by ingesting fluids, with no underlying physiologic disorder, DOT has lowered the level at which the specimen is to be considered substituted by the MRO to 2 mg/dl. Formerly, negative specimens were considered substituted with levels of creatinine between 2 and 5 mg/dl. These required the MRO to interview the donor, and in the absence of conditions that would lead the MRO to believe there was a reasonable probability that such levels of creatinine and specific gravity could be produced physiologically by the donor, verify the result as a refusal to test. These are now to be verified by the MRO as negative and dilute with the requirement that the specimen be re-collected immediately under direct observation by an observer of the same gender as the donor. This issue will continue to be controversial and additional changes to the approach to the problem of ultradilute specimens may very well be forthcoming as experience and research expand. Needless to say, any such changes are bound to affect MRO practice.

As noted above, urine has been the traditional medium or matrix for drug testing. Most workplace drug testing programs still collect urine. Currently, federally mandated alcohol testing utilizes breath or oral fluid (i.e., saliva) specimens. At present, for federally mandated drug testing only urine specimens may be collected, with the exception of instances where the donor is unable to produce sufficient urine for testing due to a permanent or long-term medical condition and a negative result is required. However, additional matrices are emerging that are gaining greater acceptance and wider use. Blood has always been available, but the invasive collection method and the greater expenses involved have relegated it to a very minor role in workplace drug testing programs. More recently, hair testing has been implemented in a number of state-sponsored (e.g., Florida) or other unregulated programs. Oral fluid testing is being offered by analytical laboratories for unregulated testing and some companies are starting to adopt it. On the horizon are other matrices, including sweat, that may also emerge as practical media for drug testing. The MRO will have to become familiar with the pitfalls and nuances of interpretation of results obtained from these alternative matrices.

The scope of workplace drug testing has largely been confined to the ten drugs or classes of drugs mentioned above. However, changes in the drug of choice of some members of industrial populations over time are being noted and calls for an expansion of the drugs tested for on regulated and unregulated testing are being increasingly made by employers and other organizations commissioning testing. At present there is pressure to include methylenedioxymethamphetamine (MDMA or "ecstasy"), gamma-hydroxybutyric acid (GHB or "Georgia home boy"), Rohypnol (flunitrazepam or "the date rape drug"), OxyContin (oxycodone), and others in standard drug testing panels. Obviously as the scope of drug testing expands this will increase the challenges faced by the MRO.

Workplace drug testing has traditionally been based on collecting a specimen at or near the workplace. The specimen is then sent to an analytical laboratory distant from the workplace. The results of the analysis have then been reported to the MRO who has reviewed them before they are released to the employer or other organization commissioning the test. Currently, federally mandated drug testing must be done in this way. However, new systems of testing have been developed that permit reading of the result immediately after collection at the workplace. In practice, most organizations performing this on-site drug testing will immediately act on negative results without any further testing, e.g., permitting a pre-placement applicant to begin work immediately, etc. Specimens demonstrating non-negative results upon on-site testing are usually sent to an analytical laboratory for confirmatory testing. If results are confirmed as non-negative, definitive action is then taken in response to the confirmatory testing done off-site. The accuracy, sensitivity, and specificity of the various on-site testing systems currently available vary, and approach the accuracy, sensitivity, and specificity of traditional analytical laboratory testing to varying degrees depending on the system used. Depending on the design of the system there may be no role at all for the MRO in the review of negative results of on-site testing, with MRO review being reserved for non-negative results that are confirmed as non-negative as described above. To what degree and in what fashion the MRO is to be integrated in on-site testing will obviously be of great interest as use of on-site testing grows.

CHAPTER 11

Quality Practices in Workplace Drug Testing

John M. Mitchell, Ph.D. and Francis M. Esposito, Ph.D.
Health Science Unit, Science and Engineering Group, RTI International, Research Triangle Park, North Carolina

CONTENTS

11.1 HISTORY

The roots of workplace drug testing lie in the U.S. military's drug testing program. The military began drug testing military personnel in response to the rise in drug abuse that accompanied the Vietnam War. Initially, it was designed to identify "at risk" individuals and to present an opportunity for treatment. By 1980, drug abuse in the U.S. military was rampant. A survey found that approximately 50% of enlisted personnel had used illicit drugs within the past 10 days.[1] It was obvious that more stringent measures were necessary. At first, the military turned to the resources available in the civilian laboratory community, but it soon recognized that the methodologies available for the testing of the large number of specimens needed to support military goals were woefully inadequate. A taskforce consisting of civilian advisors, testing industry representatives, and military personnel was organized to develop a system that would allow testing of large numbers of urine specimens in a forensically defensible manner. It was through these measures that modern-day forensic urine drug testing laboratories were created. These efforts are notable, because the military system established the basis for legal and scientific acceptance of the methods and procedures utilized in today's workplace drug testing. Important innovations in the military's program include random testing of personnel and instrumented on-site testing. This integrated program resulted in a reduction in drug use by enlisted personnel such that by 1982 a survey found that the number of personnel who used drugs in the past 30 days had dropped to less than 30%. By 1988, the number dropped to less than 10%.[2]

The military drug testing program was built upon the experience obtained from mass testing of biological samples in the clinical laboratory and forensic principles. While the primary purpose of the military's program was to maintain national security by removing personnel who were abusing drugs, it incorporated a quality system to protect service members from inaccurate results. The quality system that provided reliable testing included: observed collection of urine specimens, written standard operating procedures (SOPs), separate testing methods for initial and confirmatory tests, mass screening with immunoassay tests, confirmation of immunoassay positives by gas chromatography/mass spectrometry detector (GC/MS), utilization of deuterated analytes as internal standards for GC/MS procedures, internal and external quality control systems, independent assessment of laboratory performance by a team of experienced scientists, and the right of an accused to have a portion of the specimen retested.

Workplace programs mandated by the federal government and many state governments have incorporated most of the quality assurance components from the military's system. One component, routine observed collections, has not been incorporated into these systems, and this omission has proved to be a problem for "urine only"–based workplace programs. Unobserved collections have fostered an industry dedicated to the subversion of these programs by the provision of synthetic urine, negative human urine, prosthetic devices for the delivery of substitution products, oxidants, fixatives, peroxidases, acids, bases, and other substances that are intended to interfere with the testing methods.

11.2 THE NEED FOR QUALITY ASSURANCE

The necessity for maintaining quality in current workplace drug testing programs becomes apparent when the potential impact of this testing is considered. In 2002, laboratories certified by the Department of Health and Human Services (HHS) tested an estimated 6 million urine specimens collected from donors under federal mandate and an additional estimated 23 million specimens collected under other workplace programs (source: data supplied by federally regulated laboratories that participate in the National Laboratory Certification Program, NLCP). With estimated positive rates of 2.5% in regulated programs and 4.8% in nonregulated programs,[3] this would mean that in 2002, these laboratories reported an estimated 1 million results consistent with drug use. In order that innocent individuals do not lose their jobs, and are not denied gainful employment or placed under suspicion of drug use, the goal for quality assurance (QA) in workplace testing programs must be zero tolerance for errors. In the following sections, we review the parts of workplace drug testing programs that are critical to success, examine the programs and methods currently in place to measure the quality of the system, and provide some suggestions for future measurements.

11.3 PARTS OF THE WORKPLACE DRUG TESTING QUALITY SYSTEM

To obtain a goal of zero errors, the quality system must encompass all parts of the workplace drug testing program. The people, the equipment and instruments, the materials, the methods, and the facilities utilized in all phases of the program must be considered.

11.3.1 The Employer

In workplace drug testing, the quality system begins and ends with the employer. Without proper planning, training, and guidance, the program developed by an employer may not meet the requirements mandated by law and other guidelines. The choices an employer makes in selecting collection sites, testing facilities, and Medical Review Officers (MROs) will undoubtedly influence program effectiveness. While cost is a consideration in making these selections, it should not be the only factor considered. It is important that each test result be supported by proper collection, accurate and legally supportable testing, and appropriate determination of drug abuse. Without these elements, the actions taken by the employer as a result of a drug test may not be in accordance with the law or other guidance, may place the employer in jeopardy, and may wrongly accuse a valued employee or a qualified candidate for employment.

11.3.2 The Donor

The donor is an important part of the quality process, but the controls that can be placed on the donor are limited in most programs by considerations of privacy and personal rights. While most donors are conscientious and trustworthy, the quality system must limit the opportunity for a small drug-using minority to subvert the test by substituting their specimen with another urine or other aqueous solution, diluting their specimen by adding water or drinking large amounts of water before collection, or adulterating the specimen with substances that are meant to interfere with testing.

11.3.3 The Collection

A specimen must be properly collected to ensure an accurate test. Quality practices need to be followed to ensure the identity and integrity of the specimen. The identity of the donor must be determined, and the link between the specimen and the donor must be maintained throughout the process. Once obtained, the specimen must be secure from tampering and handled under forensic guidelines. There should be no opportunity for subversion of the drug test by the donor, the collector, or the two in collusion. These practices apply to specimens collected for on-site testing or shipped to a testing facility.

Specimens collected for workplace drug testing must be closely scrutinized for the quality of the collection. A positive drug test or an abnormal specimen validity test must be able to withstand the challenges of an MRO, a donor, or a legal review. Components of a quality collection include the collector, collection site, collection materials, and collection protocol. One source of guidance for urine specimen collection may be found in the handbook published by the Substance Abuse and Mental Health Services Administration (SAMHSA).[4] Some of the information in this section was obtained from this handbook, with a focus on the items critical to the quality of the collection process.

11.3.3.1 The Collector

A key element in a quality performance by a collector is training. A collector may be responsible for collecting one or more specimen matrices (i.e., hair, urine, sweat, and oral fluid) and may conduct on-site tests on some of these matrices. Collectors must be thoroughly trained in the collection process for each matrix. They must be trained not only for the routine collection, but also for problems that might arise during the collection (e.g., unacceptable specimen temperature, apparent adulterated specimen, insufficient amount of specimen). For collectors also conducting on-site testing, training must be specific for each on-site testing device. The training should include when and how to conduct the test, demonstration of testing proficiency, actions to take for borderline results, how to package specimens for shipment to a testing facility for additional testing, completion of documentation, and reporting of negative results.

Proper collector training begins with a qualified trainer. For urine collection, which is the most common matrix in workplace testing, a supervisor of the collection facility with previous experience typically assumes this responsibility; however, there are organizations such as the Drug and Alcohol Testing Industry Association (DATIA) that will also provide a standard course on specimen collection. With the introduction of other matrices and on-site testing into federal and state programs, all training programs may require changes. New initiatives between the testing and collection industries and controlling agencies will become necessary to ensure the validity of the specimens and the results of on-site tests. Training will need to be provided for each specimen type and each on-site test device. Training should go beyond the standard classroom lecture format. Error-free mock collections should be demonstrated for each specimen type, and testing proficiency using blind controls should be demonstrated with each type of on-site testing device. Written exams may also be part of the training program.

All training must be documented in a manner that can be easily reviewed and understood by an outside auditor. Minimally, documentation would include a description of the training, time of conduction, identification of the trainer, results of all examinations, and criteria for acceptable performance.

Beyond the initial training, a collector should be monitored for performance. Errors that occur during specimen collection or with the use of on-site testing devices require error correction training. For example, the Department of Transportation (DOT) requires collector correction training when errors in the collection process of urine specimens cause a test to be canceled.[5] The DOT requires the collector to demonstrate proficiency in the collection procedure by completing one uneventful mock collection and two mock collections related to the error. The person providing the retraining attests in writing that the mock collections were performed correctly with no errors. However, this process fails to address frequency of errors, which is an indicator of the collector's overall performance and the potential for test-canceling errors.

Within the urine drug testing industry, the collection site is perceived to be the weak link in the quality system. To correct this perception, it is recommended that the performance of all collectors be monitored and training requirements be standardized. Within federal programs, one approach to monitoring collector performance would be to examine the chain-of-custody documents, commonly referred to as custody and control forms (CCF), completed by a collector. CCFs failing to conform to federal guidelines would initiate further assessment of the collector's performance and appropriate corrective actions. Examples of collector noncompliance would be submission of a single specimen when a split specimen is required, submission of two specimens with differing physical appearance as a split specimen collection, or submission of CCFs containing multiple administrative errors. Collectors for non-federal programs might be monitored by professional groups such as DATIA.

A collector's proficiency with on-site testing devices should be measured with periodic performance testing (PT) samples for each on-site device utilized, as well as routine submission of a percentage of the on-site negatives to a laboratory conducting instrumented immunoassay testing and confirmatory testing of immunoassay positives. Abnormal numbers of false negatives should then be investigated to determine if the cause is the on-site device or the performance of the tester. Federal programs should be able to monitor on-site devices and tester performance; however, other programs may have to rely on professional groups for this function.

11.3.3.2 Collection Site

A collection site must meet certain requirements to ensure a quality collection. The site must have restricted access. Collection materials, records, and specimens must be properly stored and secured from unauthorized individuals. Materials (e.g., controls, reagents, testing devices) must be kept in acceptable temperature and humidity conditions for proper performance. Secure storage must be provided for specimens until they are tested or shipped to a testing facility, and records must be retained as required by applicable regulations. As part of its records, a collection site must maintain copies of current and past SOPs, a copy of each CCF, log books or log sheets, temperature logs from storage areas for specimens and perishable supplies, inventory of supplies/materials, donor information, and chain-of-custody handling. These records should be retained for the length of time required by governing regulations, usually 2 years or more. Urine collection sites should place bluing agent in the toilet bowl and tank and restrict access to soap, water, cleaning agents, and other materials that could be used to adulterate a specimen.

11.3.3.3 Collection Materials

A quality collection requires a collection kit designed for the matrix to be collected. The kit should contain a single-use collection device, container(s) for shipment of the specimen to a testing facility, CCFs, and tamper-evident seal(s). Kits for some matrices may require additional materials (e.g., scissors for a hair collection, wipes for a sweat collection). If on-site testing is to be conducted, the kit may also contain an on-site device. All materials must be proved not to affect the testing of the specimen. Containers used for storage/shipment must be capable of holding a tamper-evident security seal at room and frozen temperatures. Additional security can be provided to transported specimens with the use of individual sealable bags or boxes. Shipping containers can be used to protect the specimen from physical damage during transport.

11.3.3.4 Collection Protocol

A detailed written protocol (SOP) must be followed for every type of specimen collection. The basic requirements of a collection are the following:

Preparation of the collection site (discussed above)
Verification of donor identity
Preparation of donor for specimen collection (e.g., removal of coats and hats for unobserved collections)
Inspection of the specimen to ensure proper amount (and correct temperature of urine) and inspect for evidence of adulteration or substitution
Preparation of the specimen for testing, storage, or shipment (e.g., sealing the specimen with tamper-evident labels)
Completion of the CCF

The CCF is used to identify the donor, collector, employer, and MRO; to provide a unique specimen identification number; to account for handling of the specimen; and to report results and remarks in a uniform manner.

11.3.4 The Testing Facility

Testing facilities should establish QA programs to ensure the quality of their processes. Central to the quality assurance program is an SOP manual that details all procedures and processes. This manual and the documentation generated from a QA program in a workplace drug testing facility must be available for review by client auditors, certification agencies, lawyers, judges, etc. Users of the testing services in these facilities need to be assured that the results are high quality and legally defensible. A comprehensive QA program must include the handling, testing, and reporting processes. Components of QA practices include training personnel; validating and maintaining analytical instruments, analytical equipment, and computer systems; validating analytical methods; monitoring chain-of-custody procedures; using quality control samples; and participating in PT programs.

11.3.4.1 Personnel

Testing facility personnel must be well trained and motivated. They must not be allowed to succumb to the routine nature of the procedures utilized in today's high-throughput facilities. They must be knowledgeable of the scientific principles underlying the analytical procedures, the capabilities and limitations of the equipment they utilize to perform their work, and the regulatory requirements related to their work. They must be well versed in forensic principles and procedures, and their qualifications, training, and proficiency must be well documented.

11.3.4.2 Equipment and Instrumentation

The equipment and instruments utilized to test specimens are an intrinsic part of the quality system. They must be appropriate for the task and proven to perform the tasks through validation processes. This also includes the facility's management information system (MIS) and other computer-controlled equipment.

11.3.4.3 Materials

The importance of quality materials is often overlooked until a problem develops. As with equipment, the quality of the materials must be proven, not assumed. Quality facilities should perform acceptance tests on materials prior to their use and monitor their performance as a part of routine operations.

11.3.4.4 Methods

The methods utilized by a facility for all aspects of specimen handling and testing are a critical part of the quality system. A facility may have the best equipment available, highly motivated and trained personnel, and proven materials, but if the method is poorly designed, it will fail more often than a scientifically sound, operationally rugged method. Analytical methods and procedures utilized in workplace drug testing must be validated and characterized to provide timely results that will withstand the scrutiny of legal and administrative proceedings. Their continued performance at the desired level must be monitored by internal and external quality systems.

11.3.4.5 Basic Quality Assurance Practices

Method Validation

The quality of an analytical procedure is documented during its validation. Performance parameters commonly determined during validation of initial test procedures are linearity, specificity, precision, accuracy, and carryover studies. Validation of quantitative confirmatory procedures also includes sensitivity, limits of detection, and quantitation and ruggedness. Validation samples are prepared in the matrix to be tested and within the concentration range of interest. At a minimum, these tests are performed annually and when major changes (e.g., change in extraction procedure, new instrumentation) are made to an existing method. Validation records must be organized for an auditor's review and should include the purpose, scientific principle, method, results, discussion, summary, and review with approval by the facility director.

Quality Control

Quality control (QC) is a subsection of QA. QC is used to determine if all components of an analytical process are performing correctly. QC samples of known content are analyzed with the test specimens. Their purpose is to monitor the performance of an analytical procedure within defined limits of variation.

Calibrators are samples of known content by which the identification and concentration of an analyte in a specimen are determined. Calibration of a device or instrument can be established with a single calibrator or a series of calibrators at varying concentrations. It is common to find initial and confirmatory workplace drug tests evaluated with a single calibrator, but this approach requires the concentration to be determined at the administrative cutoff concentration or decision point. Multiple calibrators can be used to extend the limits of accurate quantitation on an instrument.

Calibrators and controls (often referred to as QC samples) are routinely prepared in the matrix of interest to minimize matrix effects. A quality practice is to prepare QC samples from a reference material of documented purity and content. Additionally, calibrators and controls should be prepared with reference materials from different suppliers; otherwise, a bias may occur. If both types of QC samples must be prepared from a single reference material, then they must be prepared independently.

QC samples must be validated before they are placed into use. The quantitative criteria for acceptance of calibrators should be more stringent than the criteria applied to other QC samples. Typically, calibrators should not differ by more than 10% from the target value, whereas controls may differ up to 20%. The primary method of validating new QC samples is by parallel testing with QC samples in use. If available, externally certified reference samples should be included in the validation process. The history and use of QC samples should be documented with information that includes the lot number, date of preparation and first use, expiration date, individual preparing the sample, preparation procedure, and validation data.

Workplace drug testing utilizes administrative cutoffs. QC samples at concentrations above and below the cutoff are used to demonstrate linearity around the cutoff, allowing clear differentiation of positive and negative specimens. These controls should be within the linear range of the assay and near the cutoff ($\pm 25\%$). Analytical batches should also contain a negative control to demonstrate the response of the assay in the absence of an analyte and a blind sample to demonstrate that the analytical batch has been properly prepared and analyzed. Other controls may be included to demonstrate extended linearity, lack of carryover, completeness of hydrolysis steps, lack of interference from structurally related analytes, and adherence to dilution protocols. Typically, a minimum of 5 to 10% of QC samples are required in each batch of specimens, some of which should be distributed throughout the batch.

Internal standards are essential to accurate quantitative analysis of QC samples and donor specimens. Ideally, the concentration of the internal standard should be close to the cutoff. Deuterated analogues of the analyte of interest are best for GC/MS selected ion monitoring as they have similar fragmentation patterns. However, structurally similar analytes can serve as the internal standard if the available deuterated internal standards are unsuitable (e.g., coelute with the analyte of interest and have the same major ion fragments).

Quantitative and qualitative acceptance criteria for initial and confirmatory tests must be established and adhered to during analysis. QC results, acceptable and unacceptable, must be documented as a complete record of QC performance. Systems for monitoring QC results should be established to detect shifts, trends, and biases. Actions to be taken to correct these anomalies should be contained in the facility's SOP.

11.3.4.6 Physical Plant

The physical plant in which testing occurs must be sufficient to meet requirements for security, habitability, safety, and performance of the work required. It must provide security for the specimen, the testing, and the results such that the linkage of the result with a donor is never in doubt. It must provide adequate space, ventilation, and other features to ensure a safe environment for employees and an optimum environment to conduct testing. Errors can be introduced in the testing process when the physical plant is inadequate.

It is important that an alternative source of power be available during prolonged power outages. At a minimum, there must be a plan to ensure proper temperature conditions for stored specimens.

11.3.4.7 Management Information Systems

An MIS is designed to improve the reliability, efficiency, and productivity of the testing facility. It may exist as a centralized data system with workstations tied into analytical equipment or stand-alone personal computers. It is used to log in and track samples, order tests, receive and manage data, report results, and handle routine administrative tasks. The MIS must be independently validated before being utilized for any task associated with the receipt, handling, testing, data review, or result reporting. All software and hardware added after the initial validation should be validated prior to use. Security must be in place to prevent unauthorized access, inappropriate release of information, and introduction of unauthorized, unvalidated software. The MIS must be monitored for problems, and audit trails must be established for all functions. Backup and disaster recovery procedures should be in place to prevent loss of information entered into the system.

MIS procedures should be subjected to routine audits as part of an inspection process. An MIS SOP manual describing the items above must be established and available during inspections. The person responsible for management of the MIS may not be the testing facility director; however, the director must be knowledgeable of the functions performed by the MIS, have input into its functions, and ensure that the forensic testing is supported. The testing facility director and MIS manager must work together to establish procedures for requesting changes to the MIS, validate changes, and determine end-user acceptance.

11.3.5 The Medical Review Officer

Testing facilities in all federal and many nonregulated workplace programs report test results directly to an MRO. Both DOT regulations[5] and HHS Mandatory Guidelines[6] contain requirements for MROs. To meet these requirements, physicians may obtain training and certification from any one of three organizations: American Association of Medical Review Officers, American College of Occupational and Environmental Medicine, and American Society of Addiction Medicine. To correctly interpret test results, MROs use their medical knowledge and their understanding of the applicable laws and regulations, the testing process, the collection process, and information provided by the donor. The properly trained and motivated MRO must be willing to question not only the donor, but also the collection site and the testing facility to ensure that the employer is provided a determination that considers the applicable science and regulations. MROs are often called gate-keepers because they are able to query the entire process to ensure that the quality system was intact.

The MRO practice should have a detailed SOP to ensure the quality of the MRO process, consistency of procedures, and adherence to governing guidelines and regulations. A more complete description of the MRO practice is described elsewhere in this volume.

Currently, MROs are not routinely monitored for the quality of their review. One approach to an MRO review program could be based on the documentation received by testing facilities. CCFs containing results for specimens reported as adulterated, substituted, invalid, and drug positive could be audited to determine the final disposition of the test. Reviews that revealed noncompliance with regulations would require an in-depth audit of other non-negative results reported by that MRO. Reports of these reviews could then be provided to the employer and other agencies, as appropriate.

11.4 MEASURING QUALITY IN THE WORKPLACE SYSTEM

Measuring the quality of workplace testing requires the establishment of standards against which the measurement may be performed. Currently, there are four programs that establish standards for testing facilities and conduct measurements against those standards. While there are programs that accredit or certify MROs and collectors, none conducts measurements beyond the initial certification or accreditation. Subsequent sections will describe some of the tools utilized by organizations to measure the quality of testing facilities.

11.4.1 Testing Facility Certifying and Accrediting Organizations

11.4.1.1 NLCP

The reliability of facilities that test specimens for drugs in the workplace will always be of concern. A high degree of certainty of results is required of drug testing facilities to prevent false accusations against those undergoing testing. For this reason, the "Mandatory Guidelines for Federal Workplace Drug Testing Programs"[6] was developed to provide guidance to federal agencies for the collection and analysis of urine specimens collected from federal employees. With these Guidelines, HHS established standards for certification that have made HHS-certified urine drug testing laboratories the "gold standard" in the drug testing industry. Regulations issued by the DOT for the drug testing of safety-sensitive private sector employees in its operating modes, as well as the Nuclear Regulatory Commission for fitness for duty employees, require the use of HHS-certified laboratories to perform their mandated drug testing.

The NLCP was developed for the establishment of initial and ongoing certification of forensic urine drug testing laboratories. The NLCP is designed to examine laboratories for compliance to the Mandatory Guidelines. In doing so, it examines the laboratory quality practices discussed above. Two of the major components of the NLCP are the PT program and the inspection program. Other components of the NLCP are described elsewhere in this volume.

The NLCP PT program emphasizes the laboratory's ability to accurately quantify drug concentrations in urine and report the results following Federal requirements. Strict scoring policies mean that a certified laboratory's failure to successfully test the PT samples may result in suspension or revocation of its certification. The NLCP challenges laboratories with PT samples formulated to mimic real specimens from donors in order to identify problems and to ensure appropriate actions are taken to correct those problems. Laboratories are expected to investigate all PT errors and must submit documentation of investigation and completed corrective action, as directed.

The NLCP categorizes laboratories (Category 1 — smallest to Category 5 — largest), using an objective system based in part on laboratory size, regulated specimen workload, and number of non-negative specimens reported in a 6-month period. Depending on category, semiannual maintenance inspections last 2 or 3 days and consist of one or two inspectors performing a general inspection (i.e., reviewing procedures and observing practice) and one to six inspectors performing a records audit. The NLCP focuses inspections on the procedures of the laboratory and on examination of the laboratory's forensic product through the audit component. The teams use two checklists: the General Laboratory Procedures Checklist and the Records Audit Checklist. The inspection team submits two summary reports, one from the general inspectors and one from the auditors. The inspection reports are reviewed by the NLCP technical staff. An Inspection Final Report is returned to the laboratory listing the deficiencies. Major deficiencies require a remedial action plan within 5 business days. All others must complete the remedial process within 30 days. Remedial actions are reviewed at the next inspection.

11.4.1.2 College of American Pathologists

The College of American Pathologists (CAP) has developed a Forensic Urine Drug Testing (FUDT) Accreditation Program that parallels the NLCP but is directed at non-federal workplace drug testing. Although this is a voluntary program, many clients require this accreditation for their non-federal employee drug testing.

The CAP program requires completion of a self-inspection checklist between on-site inspections that are conducted by a group of volunteers in the forensic toxicology field. Inspectors, guided by a checklist, focus on laboratory procedures and processes that include specimen handling, analytical instruments and procedures, quality control, personnel, computer operations, safety, facilities, records, and reporting.

Laboratories are also required to participate in at least the Urine Drug Testing Confirmation PT program. Acceptable performance is based on correctly identifying and quantifying drugs and obtaining a minimum score of 80%. A false-positive drug report is a survey failure with accreditation probation. Continued failures result in loss of accreditation.

The FUDT program differs from the NLCP in that the laboratory is evaluated for drugs other than the illicit drugs used in the federal program. In addition, the laboratory may use initial and confirmatory cutoffs unlike those stated in the Mandatory Guidelines. A more detailed description of the FUDT program can be found elsewhere.[7]

11.4.1.3 State of Florida

The State of Florida has established a Drug-Free Workplace program similar to the HHS Program for Federal Employees.[8] It regulates workplace testing for state employees and private workplace drug testing programs seeking discounts on Florida workers' compensation insurance premiums. Oversight of the program is maintained by the State Agency for Health Care Administration.[9] It allows the testing of biological samples (primarily urine and hair) for ten drugs/drug classes and their metabolites and alcohol (blood). It requires laboratories to conduct initial and confirmatory testing of positive specimens using two different scientific methods, to participate in proficiency testing programs, and to report test results to an MRO. Laboratories must participate in a proficiency testing program that tests for all of the required drugs. Inspections of licensed laboratories are semiannual and are conducted by the state. An HHS-certified laboratory may request to substitute one of the state inspections with an NLCP inspection and is required to submit the results of all NLCP inspections. Blind PT specimens are not required as part of the QA program.

11.4.1.4 State of New York

The State of New York amended its public health laws in 1966 to require the licensure of clinical laboratories by establishing minimum qualifications for directors and by requiring that the performance of all procedures employed by clinical laboratories meet minimum standards accepted and approved by the state department of health.[10] The laboratory licensure program is managed by the state's public health laboratory for 27 specialties of laboratory science, including forensic toxicology. The laboratory director must hold a certificate of qualification from the state department of health; minimum qualifications include a doctoral degree and 4 years of postdoctoral laboratory experience, 2 of which must be in the toxicology specialty. Laboratories engaged in the toxicological analysis of biological specimens are required to participate successfully in the state's proficiency testing program and must be in substantial compliance with standards of practice as assessed through biennial inspections. Laboratory standards and proficiency testing performance requirements in the forensic toxicology specialty are comparable to those of the NLCP. However, the State of New York has not imposed limitations on the specimen matrix, the drugs to be tested for in workplace programs, or the assay cutoff concentrations. These service characteristics are established through laboratory-client contracts. Performance in proficiency testing is evaluated in the context of the assay cutoff concentrations that are reported to the program. The proficiency testing specimen matrix is urine. Blind PT specimens are not required as part of the laboratory licensure program.

11.4.2 Methods of Measuring Quality

There are four primary systems for measuring the quality of a drug testing program: inspections, PT, blind specimens, and retests of non-negative specimens (conducted under most programs at the request of the donor).

11.4.2.1 Inspections

The inspection process is essential for improving and maintaining the quality of a testing facility, collection site, collector, and MRO. An inspection should represent an independent review of operations and be conducted by trained and knowledgeable professionals. It provides a means of peer review and feedback. Inspections provide a snapshot of the operations at the time they are conducted and are not a substitute for other quality systems. Since each part of the workplace drug testing system is dynamic, constantly changing to improve quality, increase efficiency, and reduce expenses, it is important that inspections occur on a regular basis. Although accreditation/certification may not be mandatory or available for each part of the system, it would provide additional assurance to clients that quality services and results are being provided. Currently there are no provisions for routine inspections of collection sites, collectors, and MROs.

Typically, inspections of a testing facility are a requirement for initial accreditation or certification, as are periodic maintenance inspections by certifying or accreditation organizations such as those previously described. Technically and professionally qualified inspectors receive training and continuing education sponsored by those organizations. The inspectors review all of the relevant administrative and technical functions of the site to ensure they are in accordance with prescribed regulations and industry practices. Inspectors, guided by an inspection checklist, review and evaluate testing processes to include: SOP manual, specimen handling, analytical equipment and maintenance, computer operations, analytical procedures, review and reporting of results, QCs, and personnel qualifications and training. Inspectors also review records that include specimen data, method validation data, PT data, chain-of-custody documentation, and reports transmitting results to the client or MRO. Deficiencies noted during the inspection must be corrected and reviewed at the next inspection.

A critical issue in any testing facility inspection is the amount of data that should be reviewed. The data should be carefully selected to optimize the detection of possible administrative, technical, and forensic issues. This is best accomplished by reviewing documentation and results for specimens that were reported as non-negatives. Non-negative specimens include drug positives, abnormal specimens (i.e., adulterated, diluted, and substituted), and specimens received but not tested. The resources required for an inspection should be determined by the testing facility's workload, the number of non-negative results reported, and the amount of review necessary to determine that all procedures have been adequately evaluated. Inspections meeting these criteria consume large amounts of time and resources, but are essential to maintaining quality.

11.4.2.2 *Proficiency/Performance Testing (PT)*

Participation in PT programs is an important quality practice for all testing entities, including testing facilities and personnel conducting on-site tests. They provide a mechanism for assessing the accuracy of the handling, testing, and reporting processes. They can be used to identify and test modifications to operating procedures, methods, and equipment. Regulatory agencies now require testing facilities to maintain acceptable performance in designated PT programs as a condition for initial and continued accreditation/certification. Although there are no regulatory requirements for routine PTs for on-site tests in workplace drug testing, some on-site testing facilities are using available PT programs to monitor performance.

PT samples are prepared by fortifying the matrix of interest (e.g., urine, oral fluids, hair) to the desired drug analyte concentration. Within workplace drug testing, PT samples are normally formulated to determine the ability of the testing facility to accurately identify and quantify drug analytes alone or in the presence of possible interfering substances, to identify abnormal or adulterated specimens, and to determine the dynamic range of the testing methods.

PT samples should be handled in the same manner as donor specimens. This includes not using replicate analysis if routine samples are not run in replicate. Criteria used to identify and quantify specimens should be the same for donor and PT specimens. It is ideal for a testing facility to include PT samples among donor specimens when testing the samples.

Reported PT results from participating facilities are compared to a reference value (normally the average of all results excluding outliers) to determine acceptable quantitative performance. If different analytical methods are used, means for each method are sometimes determined for comparison. Appropriate ranges for each method's mean can be standard deviations, percentage from the mean, or both. Method means can be further used to determine acceptable methods and/or instrumentation when compared to established reference methods.

The number of specimens in a PT cycle is small in comparison to the total specimens analyzed in a facility. For this reason, a quality practice is to participate in as many PT programs as possible. It is desirable to spread these PT sets over the entire year.

Testing facilities should review the information returned by the PT program. Any trends that are developing should be examined. Unacceptable results are to be investigated and corrective action taken to prevent reoccurrence. For major errors (e.g., false positives, large variations from the reference value), immediate resolution of the error through remedial action should be required by the certifying/accreditation agency.

Do PT programs make a difference in performance? Analysis of data from the NLCP PT program indicates that a PT program can enforce an initial level of performance and over time improve the performance of conforming facilities. In Figure 11.1A, bar graphs depict, through average coefficients of variation (CV), the performance of laboratories that applied for certification and certified laboratories in 1990. The variation within the population of certified laboratories was less than that of the candidate population for all drug analytes. In Figure 11.1B, it can be seen that since 1990, the average CV for most drug analytes has decreased. Not shown are the average CVs for codeine and phencyclidine, which have remained at 9 to 12% since 1990.

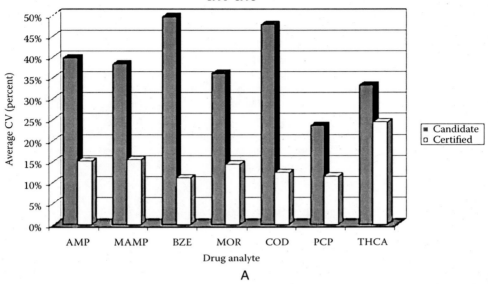

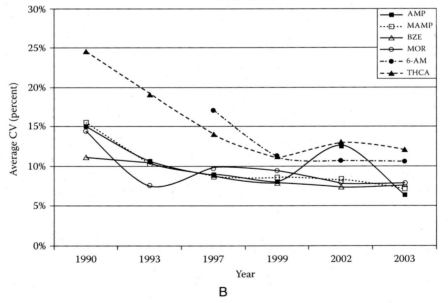

Figure 11.1 (A) Comparison of the average CV of GC/MS values reported by NLCP candidate and certified laboratories in 1990–1991 for amphetamine (AMP), methamphetamine (MAMP), benzoylecgonine (BZE), morphine (MOR), codeine (COD), phencyclidine (PCP), and 11-nor-9-carboxy THC (THCA). (B) Average CV for amphetamine (AMP), methamphetamine (MAMP), benzoylecgonine (BZE), morphine (MOR), 6-acetylmorphine (6-AM), and 11-nor-9-carboxy THC (THCA).

11.4.2.3 Blind Specimens

QA specimens presented to a testing facility in the same manner as donor specimens are referred to as blind specimens. They are submitted with fictitious information on the required custody documents and specimen bottles. These specimens should not be readily identifiable as blind specimens. The drug content of a blind specimen should be validated before use.

The HHS Mandatory Guidelines and DOT regulations require that employers submit blind specimens to certified laboratories as part of their drug testing program. Many testing facilities have QA programs that submit blind specimens to the facility as part of their quality practices. Blind specimens provide an excellent means of monitoring all of a facility's procedures. In addition to checking the technical processes of a testing facility, blind samples submitted with broken seals, incomplete CCF, etc. can also challenge the facility's procedures for handling of administrative errors.

The results of blind specimens should be closely monitored, and an investigation should be initiated when the results are inconsistent with target analyte identification or concentration. This investigation should evaluate the supplier and testing facility, identify the source of the inconsistent results, and require corrective actions.

11.4.2.4 Retests of Non-Negative Specimens

In many programs, the donor of a specimen found to be non-negative may request that an aliquot of the original specimen, in the case of a single specimen collection, or the unopened bottle, in the case of a split specimen collection, be retested. The retest is normally conducted at a testing facility different from the one reporting the non-negative result. Failure of the second testing facility to confirm the original result provides an opportunity to investigate the data from both facilities. Investigations of these incidences most often point to an issue with the collection process, usually the addition of an adulterant or a lack of a proper collection, rather than a testing facility error.

11.5 CONCLUSION

Workplace drug testing programs have become well established over the past 20-plus years. They have grown from a modest beginning in the military to become a program much larger than any imagined. The science and testing methodologies for urine drug testing have been well established. Now, it appears that the structure of workplace testing will change. The pressures from efforts to subvert a urine drug test have generated a need for solutions. Specimen validity testing cannot detect all efforts to suborn drug testing. Civil rights issues associated with observed collections make it clear that they can only be used under specific circumstances. Workplace drug testing using matrices such as hair, oral fluids, and sweat are being offered as a solution. While it would appear that these matrices are less subject to undetectable attempts to circumvent a valid test, there is still discussion within the scientific and legal communities about interpretation of results from these tests and their appropriate use. It is believed that these issues will be clarified and in the future the effectiveness of workplace drug testing will be enhanced as a result of their implementation. However, as a word of caution, this is possible only if the lessons learned about the quality requirements of the urine drug testing system are appropriately applied to the testing of new matrices.

ACKNOWLEDGMENTS

The authors express their appreciation to Donna M. Bush, Ph.D., Walter F. Vogl, Ph.D., and Charles Lodico, M.S., of the Division of Workplace Programs, Center for Substance Abuse Prevention, Substance Abuse and Mental Health Services Administration, U.S. Department of Health and Human Services; Richard W. Jenny, Ph.D., of the State of New York Department of Health; and Patricia L. James, B.S., M.A., the State of Florida Agency for Health Care Administration for regulatory insight.

REFERENCES

1. Department of Defense. Urinalysis test results analysis. Naval Military Personnel Command Contract N00600-82-D-2956, October 22, 1982.
2. Brey, R.M. et al. Highlights of the 1988 worldwide survey of substance abuse and health behaviors among military personnel. Department of Defense Contract MDA903-87-C-0854, 1989.
3. Quest Diagnostics, Inc. 2003 drug testing index. http://www.questdiagnostics.com/.
4. Urine Specimen Collection Handbook for the New Federal Drug Testing Custody and Control Form. Division of Workplace Programs, Substance Abuse and Mental Health Services Administration, Department of Health and Human Services. http://www.drugfreeworkplace.gov/DrugTesting/SpecimenCollection/UrnSpcmnHndbk.html.
5. Procedures for Transportation Workplace Drug and Alcohol Testing Programs, 49 CFR Part 40, Department of Transportation, 65 *Federal Register* 79462, December 19, 2000 and 65 *Federal Register* 41944, August 9, 2001.
6. Mandatory Guidelines for Federal Workplace Drug Testing Programs, Substance Abuse and Mental Health Services Administration, Department of Health and Human Services, 59 *Federal Register* 29908, June 9, 1994.
7. Baenziger, J. The College of American Pathologists voluntary laboratory accreditation program, in *Drug Abuse Handbook*, Karch, S.B., Ed., CRC Press, Boca Raton, FL, 1998, chap. 10.2.2.
8. Florida Statutes Section 112.0455.
9. Florida Administrative Code Chapter 59A-24.
10. New York State Public Health Law Article 5, Title V.

CHAPTER 12

Legal Issues in Workplace Drug Testing

Theodore F. Shults, J.D., M.S.
Chairman, American Association of Medical Review Officers, Research Triangle Park, North Carolina

CONTENTS

Over the 20-year life of modern workplace drug testing, a body of law has been created that is directly related to and directly influences its practice. It is an expansive body, stretching from the constitutional limits on the government's ability to require testing under the Fourth Amendment; it involves a large and expanding matrix of state drug testing laws and legal cases, and travels all the way down to local rules establishing how to introduce drug test results in an unemployment security case.[1] Over this timeframe, fundamental federal laws such as the Americans with Disabilities Act, the Omnibus Transportation Employee Testing Act of 1991, and more recently the Health Insurance Portability and Accountability Act (HIPAA) came into existence. All of these laws affect and shape drug testing practices, the drug testing industry, employment practices, and social policy.

There are a number of legal treatises that deal with employment drug testing and that cover federal and state drug testing laws. This chapter focuses on current key legal issues with a particular perspective of a drug testing service provider and employer. Regardless of the reader's knowledge of and facility with the specific laws of workplace drug testing, the following are broad issues that all employers and providers of drug testing services such as collectors, laboratories, third-party administrators, and medical review officers (MROs) will be dealing with directly in the years ahead.

12.1 LIABILITY OF DRUG TESTING

Over these 20 years or so of modern workplace drug testing there have been literally millions of drug tests performed, tens of thousands of positive tests reported, and relatively few lawsuits. There are many reasons for this, but perhaps the most significant is that the analytical procedures are reliable and defendable. Mistakes happen, and mistakes have happened, but there is a very low incidence of error considering the large number of tests performed. Other historical factors contributing to the relatively low level of litigation are that it is often difficult for a donor to bring an action against an employer or a service agent alleging negligence, and the damages of such alleged negligence are often low. One of the great concerns of drug testing has been the fact that often the only evidence of drug use is the laboratory test, and for many donors it is easier to deny illicit drug use and claim the results are a "false positive" than to acquiesce to the results and reality. Thus, there has always been a great deal of potential for litigation, which would cripple drug testing programs.

From the perspective of an employer, the greatest liability is not from drug testing but rather from the damages caused by an impaired employee to a third party. If an employee, acting within the scope of employment, causes injury to customers or to the public after consuming alcohol or drugs, liability may be imputed to the employer under the doctrine of *respondeat superior*. This vicarious liability can occur in commonplace situations, such as an accident caused by an impaired employee driving a company vehicle.[2] The actual costs of a drug- or alcohol-related accident can be astronomical, e.g., the *Exxon Valdez* environmental catastrophe and damage awards. Employers involved in hazardous or safety-sensitive work have understandable concern over the drug use, fitness, and qualifications of their employees.

One of the greatest misconceptions in drug testing is that employers who drug-test are at a great risk for lawsuits from employees over the results and adverse employment consequences. That's not true — or at least, it is an overstatement. Certainly, if an employer discriminates against an employee or applicant there are legal consequences, but courts do not want to get into routine hiring and firing decisions, or second-guessing business decisions. Thus, there is the judicial doctrine of *employment-at-will,* which broadly protects employers from direct suits from employees over drug testing.

Essentially, the employment-at-will doctrine stands for the premise that in the absence of an employment contract there is no "temporal" relationship between employer and employee. In other words, the employer–employee relationship can be terminated without notice, and without cause. Since there is no need for a reason to terminate the relationship, it is irrelevant whether the reason for the termination is reasonable, unreasonable, fair, unfair, or right or wrong, as long as it is not illegal or against public policy. (An illegal reason is discrimination based on race, age, or sex, retaliation for filing a workers' compensation claim, or termination for certain types of "whistle-blowing.") Thus, in the absence of any state law to the contrary, employers have a great deal of insulation from lawsuits arising from employment decisions based on drug test results, even if the results are not accurate.

An important distinction is when there is a collective bargaining agreement (CBA). This is a type of employment contract. The key distinction between a CBA and employment-at-will is that an employer cannot terminate a covered employee unless there is a reason, which is usually phrased as "just cause."[3] With unionized employees, the union has a duty to defend its members, and there is a right to a grievance process and arbitration. The employer may have to reinstate a terminated employee following arbitration, and pay back wages, but there are no monetary damages.

From the service provider's perspective, liability issues are more complex. First, there is contractual liability, which is the service provider's responsibility for performing drug testing services appropriately for the client (the employer). For the laboratory, it involves the accurate handling, analysis, and reporting of specimens. For collectors, it is the proper performance of the collection process. For MROs, it is fulfilling the verification process correctly. But this contractual liability exists between the service providers and the client (i.e., the parties to the contract). There have been a few situations where employers have sought damages against a service provider for breach of contract, but this has been relatively rare.

Second, the doctrine of employment-at-will has not been extended to cover service agents, even though they are acting in many cases as the "agent" of an employer. What has historically protected service agents from lawsuits from donors is the fact that the law has not recognized a duty of care existing between the third-party service agent and the donor of a specimen. Historically, the service provider did not owe a "legal duty" to the donor. Thus, a donor who alleged that the specimen was mishandled or was a "false positive" could not bring a negligence claim against the service provider, because an essential element of negligence law is the existence of a legal duty. To be a bit more accurate, the donor *could* bring a suit against the service provider, but it would be quickly dismissed. That has been changing on a state-by-state basis since the mid-1990s. That historical protection or legal insulation is eroding, and has been eliminated in a number of jurisdictions.

A good illustration of how the law looks differently at employers and service providers is the case of *Jane Doe v. SmithKline*. In this 1994 case, the Quaker Oats Company made an offer to an applicant (Jane Doe) for a marketing job. The job offer was contingent on passing a drug test. The drug test was reported to Quaker as positive for a low level of morphine. Apparently there was no MRO verification, and no clinical evidence of abuse. When it learned that the laboratory had identified morphine in the specimen, Quaker simply withdrew the offer.[4] The applicant, who had quit her job and was in the process of relocating, brought suit against Quaker and the laboratory. Even though Quaker did not use an MRO and withdrew the offer of employment, the court quickly dismissed Quaker as a defendant under the employment-at-will doctrine. It was not so easy, however, for the lab. The trial court found that it was liable.

The Texas Court of Appeals ruled that, given these facts, a laboratory had a legal duty to warn test subjects of possible influences on results (e.g., poppy seed ingestion). The Texas Supreme Court, however, overruled this decision. The laboratory analysis was correct — there was morphine in the urine. That was what the laboratory was asked to determine, and it did. In the end, a divided court found that the laboratory has no legal duty to warn donors.[5]

The failure to mention poppy seeds to the employer, and the employer's reliance on the test results, may still be a basis for a suit alleging willful and intentional interference with the conditional offer of employment between the employer and employee.

No wrong goes without a right, as plaintiffs' lawyers like to say. Thus, with employment-at-will protecting the employer and with no legal duty attaching to the provider, the donor might not have any legal recourse. Naturally, with drug testing an almost universal employment practice, this insulation was not going to last forever. The law adapts, and it is in that process.

Over the past few years there has been a case-by-case, jurisdiction-by-jurisdiction expansion of what is essentially the scope of legal liability to cover various drug test providers. There is an expanding body of law that allows employees to bring direct actions against laboratories (and theoretically against MROs, third-party administrators, and specimen collectors).

In the first significant case, *Stinson v. Physicians Immediate Care, Limited*, Stinson alleged that the laboratory had been negligent in performing a drug test on him and had reported a false-positive result for cocaine. Stinson alleged that the test result was false or, in the alternative, the report of the test result was false. As is characteristic in this type of case, the allegations were general in nature. The plaintiff alleged the defendant was negligent.

The trial court dismissed Stinson's case on the grounds that a laboratory does not have any legal duty to the donor of the specimen. The appellate court in Illinois, however, reversed this decision and held that a drug testing laboratory owes a duty of reasonable care to persons whose specimens it tests for employers or prospective employers.

The appellate court view was that it is reasonably foreseeable that the tested person will be harmed if the laboratory negligently reports the test results to the employer, that the laboratory is in the best position to guard against injury, and that the laboratory is better able to bear the burden financially than an individual who is wrongly maligned by a false-positive report.

Thus a terminated employee, and presumably a frustrated applicant, has a basis to state a claim against a laboratory by alleging that the laboratory has a duty to the donor to act with reasonable care in collecting, handling, and testing the specimen, and reporting the results accurately. Several appellate decisions have defined the legal responsibility of laboratories.[6]

This expansion of legal duty is not limited to the laboratories. In November 1999, the Wyoming Supreme Court cited the *Stinson* case when it held not only that a collector (and a collection company) owes a duty of care to the donor, but that it also has a duty of care as a consultant, where the collection company recommended a urine alcohol test. (And this duty to the donor essentially extends to all of the service providers.)

The Wyoming case was *Duncan v. Afton, Inc.*[7] It is another example of a court finding that a drug test provider (in this case, a third-party administrator, a collector, a laboratory, and an MRO) owes a duty of care when collecting, handling, and processing urine specimens for the purpose of performing substance abuse testing. This duty is not only to the employer, but to employees and applicants as well.

All of the above cases deal with the duty of drug and alcohol service providers in regard to allegations of "false positives" and protecting the donor's interests. A question that has been in the background is whether service providers such as MROs owe a duty to third parties for "false negatives" or "undue delay" in reporting results, or the failure to notify an employer that an employee may be unfit for a safety-sensitive position. In other words, is there exposure for a service provider who fails to report a "positive" test for whatever reason, and there is a drug-related catastrophic accident?

Such circumstances are rare, but one court case has dealt with such an occurrence. The case, *Turley v. Taylor Clinic,* went to trial in late 2005 in the Aniston, Alabama state court, the first reported case of a drug-related catastrophic accident occurring while the MRO verification process was under way. It is the first case where an MRO has been charged with causing harm to a "third party" — not the donor, but individuals harmed by the donor (a drug-impaired truck driver).

The case illustrates the relatively indefensible position the MRO and the MRO's employer, Taylor Clinic, were placed in by the failure of the regulations to address a fundamental flaw in drug testing analysis and the intrinsic regulatory conflict between "safety" and the prohibition of removing a presumptive positive donor from safety-sensitive tasks until the verification process is complete.

Methamphetamine exists in two forms called isomers, with significantly different clinical and pharmacological properties. Both forms look identical to a mass spectrometer. The *l* isomer, levmetamfetamine, is found in over-the-counter inhalers like Vicks Inhaler®. The *d* isomer is the form of methamphetamine that acts as a potent CNS stimulant and is the form that is a controlled substance and is abused. The *l* and *d* forms of methamphetamine can be differentiated analytically, but there is no regulatory requirement to do so. Therefore, most laboratories that must compete on price in this mandatory testing program do not do the analysis.

The risk to all MROs is that if they do not order the *d* and *l* test for a methamphetamine positive (in cases where there is no other medical explanation), the MRO is exposed to a future claim from the donor that the results were in fact due to legal use of a nasal inhaler.[8] The positive methamphetamine donor will allege that the MRO negligently or intentionally failed to order the isomer identification analysis. Thus, under the current regulatory scheme the laboratory's report of undifferentiated methamphetamine presents MROs with the Hobson's choice of either verifying a methamphetamine-positive without objective laboratory evidence of whether the drug reported is *d* methamphetamine or *l* levmetamfetamine, or delaying the verification to perform the *d* and *l* testing.

A regulatory conflict exists in addition to this technical deficiency. DOT regulations place a legal duty on MROs to remove donors who present safety risks, and yet also prohibit removal of a donor with a positive test until the result is verified by the MRO.[9]

What happened in the *Turley* case is that the laboratory reported a positive undifferentiated methamphetamine to the MRO. The MRO immediately contacted the truck driver. The truck driver first claimed that the result was due to prescription use of Didrex®, then to Adipex®.[10] He strongly denied use of illegal methamphetamine. The MRO determined that there was no prescription use and ordered the *d* and *l* analysis. The MRO promptly contacted the employer representative as soon as the laboratory analysis came back as *d* methamphetamine.

While the MRO was waiting for the *d* and *l* analysis, the truck driver was seen driving his rig erratically on a major highway. Before the state police could intercept him, the driver exited the highway and ran the stop sign at the end of the exit ramp. A pickup truck collided with the side of his tractor-trailer, resulting in the immediate death of the passenger and the subsequent death of the driver. The truck driver was uninjured.

The police arrested the driver, who could not stay awake. A police-requested blood test for drugs and alcohol was positive for methamphetamine. The truck driver was found to have grossly overextended his time on the road. He had doctored his logbooks and was suffering from fatigue. He pled guilty to vehicular homicide and was given an active jail sentence.

In the discovery process of the wrongful death suits filed against the trucking firm, the plaintiffs' attorneys discovered that the random DOT drug test for methamphetamine had not been reported to the employer at the time of the accident. To the plaintiffs, this simply looked like negligence. The plaintiffs alleged in their respective complaints that the MRO should have contacted the employer and had the driver removed the moment the lab result came in (although this would have violated DOT's regulations), and subsequently that the MRO should not have ordered the *d* and *l* isomer test, which caused a delay in reporting the result.[11]

What the plaintiffs presented at trial were the bad acts of the truck driver and his guilty plea for drug-related vehicular homicide. They pointed out to the jury that DOT regulations refer to the MRO as the "gatekeeper" for the program. This was creatively interpreted to mean the MRO is supposed to protect the public from illegal drug users — specifically, from methamphetamine-crazed truckers. The plaintiffs also read the provisions from Part 40 that state the donor has the "burden of proof" as meaning that if on the first telephone call the donor cannot prove he has an alternative medical explanation, he must be found to be positive and/or removed from service.

It mattered little that these were fundamental misinterpretations of the regulations. It did not seem to matter that the MRO role is also to protect individuals who are not illegal drug users from being falsely accused or losing employment because they use an over-the-counter inhaler. It was quite simple for the plaintiffs: local fellows are dead, an out-of-town doctor failed to report the results — bad outcome, bad MRO.

These issues could have been addressed. However, the defense was essentially forced to settle this case by following the informal guidance (or as it was described, the bombshell) that was dropped by DOT's Office of Drug and Alcohol Policy and Compliance (ODAPC). The defense counsel for Taylor Clinic asked ODAPC its view on the actions of the MRO and presented the office with a "hypothetical" synopsis of the facts. The defense counsel was surprised to learn from ODAPC that:

> … the MRO had no compelling medical reason to order the d,l isomer test and delay reporting. The donor offered no medication or explanation in the hypothetical that gave a reason for delaying the final verification decision waiting for that particular additional test.

The new concept of requiring a *compelling medical reason* to order a *d* and *l* test is without any regulatory basis and is in frank conflict with the guidance DOT has given for over a decade.

It also raises the question of whether an MRO, in the absence of a compelling medical reason, should delay reporting of *any* positive results where the purported explanation is unlikely. For example, it is common for a donor to claim that he or she is positive for cocaine because his or her dentist used cocaine in a recent procedure. It is generally understood in the MRO community, the general medical community, and the dental community that cocaine is not used in routine dental procedures in the U.S. The standard of practice has been, however, for the MRO to give the donor some reasonable amount of time to get information from the dentist. Some MROs even contact the dentist directly to verify the information. If there is a drug-related accident or injury while this aspect of the verification process is under way, can it now be alleged that the MRO was negligent because there was *no compelling medical reason* to verify the donor's claim? DOT has certainly opened the door for such allegations.

ODAPC further exacerbated its guidance to Taylor Clinic in respect to verification of methamphetamine by subsequently proposing that MROs go ahead and report positive methamphetamine results without the *d* and *l* analysis. The MRO is then free to decide whether to order the additional test. In the event the *d* and *l* results are reported back as all *l* isomer, the MRO can issue a "revised" report. This sounds reasonable, except that in most cases the donor will have already been fired, and most employers who are not governed by collective bargaining agreements will be under no legal obligation to rehire them. It is doubtful that this "guidance" is even constitutional, as it is simply an unreasonable rule with unreasonable results. At the time of this writing, these concerns have been expressed to DOT without satisfactory response.

Turley v. Taylor Clinic is the worst-case scenario for an MRO and for the government. It all stems from the failure of the HHS Mandatory Guidelines to require laboratories to identify which form of methamphetamine they are reporting. If the trend is toward holding the service provider liable for the shortcomings of policy and regulatory schizophrenia, MRO verification of DOT results will begin to present unacceptable levels of risk. Some legislative relief would be appropriate.

12.2 SPECIMEN VALIDITY: MANAGING THE INTEGRITY OF SPECIMENS AND THE QUESTION OF FEDERAL PREEMPTION

Workplace drug testing has historically been based on the analysis of urine specimens. There are many advantages to using urine. There is sufficient quantity, it is easy for a laboratory to handle and manipulate the specimen, urine is mostly water, and the physiologic processes and kidneys provide a first pass at concentrating the metabolites and drugs of interest in the urine. The development of high-speed automated immunoassay instruments in the 1980s facilitated the ability to quickly and effectively screen urine specimens and made large-scale urinalysis programs practical. Urine gives a 2- to 3-day window of detection using typical workplace drug testing cutoff levels.

The disadvantage of urine is, frankly, that it is urine. Urine collection involves a personal and private process that requires a reasonable degree of privacy. A legal (and social) constraint of urinalysis testing is the premise that in the absence of individualized suspicion of adulteration or other subterfuge of the collection process, it is unreasonable to witness or directly observe the production of a urine specimen.[12] This "reasonable" degree of privacy in the collection process has provided a reasonable opportunity for a drug user to adulterate, substitute, or otherwise attempt to undermine the drug testing process, and many have. Adulteration and substitution have challenged the integrity, technology, and economics of urine drug testing. The very principles and legal foundation of drug testing have been challenged by the adulteration industry and their willing accomplices: drug users who have no interest in changing their behavior.

In 2004, HHS implemented a technical strategy that required certified laboratories to screen all specimens received from federal agencies for the presence of oxidants, check the pH, and determine the level of creatinine.[13] The new standard is known as the specimen validity testing (SVT) rule. It establishes a technical protocol that provides a level of screening and confirmation for adulterants and the identification of substituted specimens equivalent to the identification of prohibited drugs and metabolites.

The rule was framed as the ultimate solution for the war on adulterants. However, in developing its SVT rule, HHS made a fundamental policy decision to require only a generic screening for adulterants, and not to require any laboratory to have a confirmation procedure for the screen. A survey of laboratories in 2005 indicated that the certified laboratories have essentially abandoned the process of adulterant confirmation.[14] One presumes that the extent of the laboratory retreat from this area was unanticipated, but the consequence has been the precipitous drop in the number of adulterated specimen reports.

A specimen that meets the screening criteria for a tampered or suspect specimen is then reported as an "invalid" result. Following the implementation of the SVT rule, all adulterated specimens became invalids. These "invalid" results trigger requirements for the MRO to have discussions with the laboratory, call the donor to discuss the results, and in most every case call the donor back for an observed recollection. In addition, now the MRO must even "verify" that the observed collection was indeed observed.

This approach begs the question: What are the results of these second observed collections? How many are positive? No systematic data have been released to support the fundamental premise of the SVT rule. Is the return visit of the donor with an invalid any better than requiring a second drug test of anyone for no reason? Under this rule, the mandatory recollection of specimens under observed conditions is not productive.

Unfortunately, the HHS specimen validity testing rule has been implemented and serves as a standard and template for testing federal agency employees. It has become the *de facto* standard, being authorized under DOT and NRC for regulated industry. In late 2005, DOT formally proposed adopting these rules without modification.[15] However, DOT and NRC are still evaluating the merits of the overall approach, and may ultimately decide not to follow this protocol.

A recurring pattern of specimen validity testing has been that when laboratories begin to test for a particular adulterant, such testing has a profound impact on the use of the identified adulterant. In the 1990s glutaraldehyde was one of the more common urine adulterants. Its use dropped off sharply and in a relatively short period of time after the laboratories began to identify the compound. Unfortunately, this lesson has not been incorporated into the HHS SVT rule. There are now no more confirmed adulterated test results, only invalids.

Current thinking is that alternative tests such as oral fluid, sweat patches, and even hair testing provide a better specimen because the collection is essentially one that is observed. But the adulteration industry sees these "*alternative*" specimens merely as new markets and opportunities ahead. Adulteration products are already available to "clean toxins" out of hair. One can foresee a line of products to "detoxify" the mouth and breath and to purify sweat. Like so many products in this genre, the product does not have to work to be successfully sold in this market.

To date, the legal issues of specimen validity testing have not involved the question of whether it is appropriate to test for adulterants or for substitution. The focus has been on the integrity of the analysis and the validity of the cutoff levels.

Although laboratories are not actually performing confirmation testing of adulterants, they are reporting creatinine and specific gravity results. The creatinine cutoff level for a substituted specimen was initially established by HHS at 5 ng/ml. This was, however, found to be high enough for some individuals to achieve. In other words, it was not medically or physiologically impossible to produce a "substituted" specimen at or around this level. Although a donor could challenge the substituted results and engage what was called a "referral physician" and demonstrate their ability to produce creatinine at such low levels, it was a burdensome process.

In February 2003, a colloquium funded by Congress and hosted by the FAA Civil Aerospace Medical Institute (CAMI) specifically convened to assess the soundness of the specimen validity rules in effect at that time.[16] One result of the research and findings of this colloquium was the decision to lower the creatinine cutoff level for substituted specimens.

The change from 5 ng/ml creatinine to <2.0 ng/ml was made officially by DOT on May 28, 2003. DOT also directed that specimens that had been previously reported as substituted and had a creatinine between 5 and 2 were to be canceled. They are now reported as "dilute" and require a subsequent collection under direct observation.

A separate but related concern to determining the appropriate physiological cutoff values for creatinine and specific gravity is the issue of how accurately these can be measured. A lot was learned the hard way. In the early days of testing some of the analytical instruments designed for measuring creatinine levels had a feature of truncating results, or rounding down to the lower integer. This feature was buried in the software and no one really cared about the decimal place, until substitution became an issue. Dropping the fractional number turned out to have significant consequences.

The case that received the most attention in respect to truncating creatinine results was *Ishikawa v. Delta Air Lines, and LabOne, Inc.* On July 3, 2001, a federal district court jury in Portland, Oregon, found LabOne negligent under the state negligence law in the laboratory analysis of Yasuko Ishikawa's drug test. LabOne had reported Ms. Ishikawa's specimen as substituted to her employer, Delta Air Lines, and Delta had fired Ms. Ishikawa, who worked as a flight attendant. Her dismissal was based on Delta's policy to terminate employees who have been reported as having substituted or adulterated specimens. The laboratory found itself in the difficult position of not being able to prove conclusively that Ms. Ishikawa's urine did in fact meet the existing definition of a substituted specimen, primarily because of the software truncation of creatinine results. The laboratory was also placed in the difficult position of defending a federal standard for creatinine levels that was subsequently lowered.

In the end, Ms. Ishikawa was awarded $400,000 in compensatory damages. This verdict was appealed on the grounds that federal law, in particular the Federal Omnibus Employee Testing Act, which is the current statutory authority for the DOT regulations, "preempts" state law, including negligence law. The trial court did not accept that argument, and the case was appealed to the Ninth Circuit Court of Appeals. The Ninth Circuit took the opposite view and noted that if anything, Part 40 *preserved* the donor's right to sue. The court noted that the DOT regulations expressly provide: *"[t]he employee may not be required to waive liability with respect to negligence on the part of any person participating in the collection, handling, or analysis of the specimen."*[17,18]

The court deduced that since negligence is a state common law tort, it would make no sense for the DOT regulation to prohibit requiring the employee to waive negligence claims if those claims were preempted and could not be made. This regulation prohibiting required waivers of negligence claims implies that such claims exist and are not preempted.

On November 30, 2004, the Eighth Circuit followed this decision in *Chapman v. LabOne* and *Howell v. LabOne.* These two cases were consolidated on appeal. In both cases the federal district court judges ruled that the state negligence action against the laboratory was preempted. The district court in Iowa dismissed the *Chapman* case on the ground that the common law claims were preempted by the Federal Railroad Safety Act (FRSA), as amended by the Federal Omnibus Transportation Employee Testing Act of 1991 (FOTETA), and the Railway Labor Act (RLA). The district court in Nebraska ruled that removal of the *Howell* case was proper based on the doctrine of "complete preemption" under the FRSA and the FOTETA, and then dismissed the case on the ground that the FOTETA provided no private right of action.

The Eighth Circuit reversed both of these cases and noted:

> We agree with the Ninth Circuit that "[n]egligence is a state common law tort, and it would make no sense for the regulation to prohibit requiring the employee to waive negligence claims if those claims were preempted and could not be made."

This appears to be the current majority view, which essentially means service agents (not just laboratories) performing drug testing under the DOT program will be subject to state negligence law in respect to their work, in jurisdictions that have expanded the "duty of care" to donors. The legacy of the litigation surrounding specimen validity issues has resulted in a more rugged technological approach and numerous additional safeguards for specimen validity testing. It may also be a turning point in how the industry views the litigation risks of this endeavor.

12.3 2005 CONGRESSIONAL INTEREST IN ADULTERATION AND SUBSTITUTION

On May 17, 2005, the House Energy and Commerce Oversight and Investigations Subcommittee held a public hearing to investigate the issue of the subversion of federal drug testing by adulteration and substitution. There had been growing interest in getting Congress involved in this problem, specifically seeking federal legislation to help curb the marketing and use of devices and chemicals designed to defeat drug tests.

The hearing came coincidentally on the heels of a widely covered story of Minnesota Vikings running back Onterrio Smith, who was detained by airport security who were curious about the freeze-dried urine in his carry-on luggage. No doubt they were also curious about the plastic prosthetic device that was also present. What Mr. Smith had was *The Whizzinator*, which has been on the market for a number of years. The Whizzinator consists of a prosthetic penis that will deliver a liquid at about 98.6°F. It is now sold with freeze-dried urine.

Thus, the Whizzinator became the media breakout product for the adulteration and substitution industry. But the Whizzinator is just one of approximately 400 products available today to defeat drug screening tests, according to the report by the U.S. Government Accountability Office (GAO) released at the hearing.

> The sheer number of these products, and the ease with which they are marketed and distributed through the Internet, present formidable obstacles to the integrity of the drug testing process.[19]

The GAO report was released as three major manufacturers and sellers of the adulterants and devices pleaded the Fifth Amendment, declining to answer questions at a hearing after receiving subpoenas from the House Energy and Commerce Oversight and Investigations Subcommittee.

At this time a bill has been introduced in the House of Representatives titled: The Drug Testing Integrity Act of 2005. The purpose of the bill is "To prohibit the manufacture, marketing, sale, or shipment in interstate commerce of products designed to assist in defrauding a drug test." It is a short bill that focuses on the manufacturers and distributors of these products, which are riper targets than users. Such federal legislation will not eliminate the problems of adulteration and substitution, but it will certainly help.

12.4 EXPANDING TECHNOLOGY: THE OPPORTUNITY AND CHALLENGE OF HAIR, ORAL FLUID, AND SWEAT TESTING

There has been long-standing interest and research in testing other biological specimens, such as saliva (now referred to more accurately as oral fluid), hair, and sweat. One generic advantage of all these methods is that collection of the specimens can be witnessed by a collector. A witnessed collection greatly reduces the opportunity for adulteration or substitution. There are other advantages and disadvantages of these methods, and a somewhat new legal frontier is ahead.

In April 2004, HHS published a proposed rule to amend the current Mandatory Guidelines for Federal Workplace Drug Testing.[20] The amendment would have expanded federally mandated drug testing to include the collection and testing of hair, oral fluids, and sweat. It also would have expanded urine testing to include on-site testing, decentralized screening of urine, the testing of designer drugs, and lower cutoffs for amphetamine and cocaine.

Oral fluid is relatively easy to obtain from a donor. In what is referred to as laboratory-based oral fluid testing, a stick with an absorbent pad is placed in the mouth for a short time.[21] In one system, the device is sent to the laboratory for screening and confirmation. There is more and more evidence that oral fluids perform equally as well as current urine testing in terms of prevalence rates of positive drug tests. It is a very promising technology that is particularly attractive not only to employers, but also to school districts interested in drug testing students. On the other hand, there is some concern that the method has a relatively shorter detection time for marijuana use and paradoxically is more susceptible to "passive" or secondhand smoke.

Hair testing has been a controversial area of drug testing for more than a decade. It is clear that drugs and metabolites can be sequestered in the matrix of the hair and may be identified years after use. In workplace programs, only a relatively short section from the scalp is tested. This practice is to correlate the detection window to the past 90 days of hair growth, which will fall within the bounds of the "current use" definition of the Americans with Disabilities Act. One issue that remains somewhat unresolved is how well the laboratories can distinguish use of the parent drug from environmental contamination. Some laboratories are quite adamant that this is not an issue; other experts are not so sure.

Another complexity of hair testing is that there is large body of evidence that the sequestered amount of some drugs such as THC and cocaine is proportional to the darkness of hair. There is a good correlation of the amount of drug in hair and the amount of melanin. There is a somewhat charged debate in the toxicology community, and moving into the legal community, as to the significance of this color bias in respect to its possible disparate impact on African-Americans and Hispanics.

The use of sweat is not common in workplace drug testing programs because it involves the adherence of a sweat patch to donors for a period of at least a week. Its use is almost exclusively in monitoring treatment and abstinence.

The HHS proposed rule summarized 8 years of work by industry working groups and framed many of the critical issues. It was subject to public comment and was extensively critiqued as raising as many questions as answers. The 283 formal comments raised 2000 separate issues that needed to be addressed. In 2006, the Office of Management and Budget (OMB) announced on its list of 2006 completed rulemaking that these proposed amendments have been withdrawn.

Regardless of this misadventure, much was learned. These alternative technologies will continue to develop and grow in the private sector. Private employers are still faced with the challenge of assessing the technical and practical merits of each alternative technology.

Conceptually, the alternative biological specimens provide different types of information about drug use and drug exposure. An individual's current mental and physical status and recent use may be best assessed in terms of oral fluid testing results. Past history and pattern of use may be best revealed with hair. It is unlikely that a hair test would be used for a post-accident test, simply because hair tests do not typically reflect drug use in the past few days, and there is no peer-reviewed data showing a correlation between hair concentration and behavioral effects. Similarly, the use of an oral fluid test, which provides a relatively narrow window of detection of recent drug use, would be of less value than urine in assessing abstinence for the purposes of a return-to-duty test. Urine testing has been used for all types of drug tests: random, pre-employment, post-accident, for-cause testing, and return-to-duty. It is unlikely that any of the alternative tests would have such broad application.

In the years ahead, the courts will be dealing with all of these technologies in a variety of circumstances. In general, the legal issues and litigation will be similar to what has been seen with urine testing. Is the test accurate? Was the test, collection, or interpretation done correctly? Does the employer have the right to use these tests? Do the tests violate existing state privacy or drug testing law? Do the tests and the employer's policy violate the Americans with Disabilities Act? Do the tests discriminate against a protected class?

To the degree that alternative testing methods mimic the technical safeguards in urine testing, such as using approved and accepted screening and confirmation methods and having MRO verification, the legal cases should follow the same results that urine testing has produced.

The courts (and perhaps legislatures) may also be dealing with a relatively new issue that did not exist in a one-test universe of urinalysis. The new issue is whether the selection of a particular method is "fair." Is it fair to screen some employees with urine testing and others with hair? Or is it fair to have oral fluid testing for some post-accident testing and urine for others? Does the less-rigorous standard open the door for charging that an employer is not doing an adequate job of screening employees?

Employees have a strong sentiment that testing should be fair, although their definition of "fair" may not necessarily be rational. For example, employees may feel that for drug testing to be fair everyone has to be tested at the same time, or that employers have an obligation to provide assistance and/or treatment for substance abuse. Employers also have an interest in projecting fairness. The reality is, however, that life is not fair, and there is limited legal recourse for "fairness" for employees. Here is where state legislatures have stepped in — for better or worse, depending on one's perspective.

At the federal and state level, a more rigorous constitutional test is required when a government agency performs the tests or requires an employer to test as part of a regulation or law. It is premature to speculate on the legal significance of "fairness" in the federal program, other than to say that the U.S. Constitution requires reasonableness. At some point, *fair* and *reasonable* intersect.

The constitutional requirement of governmentally mandated testing also requires a showing of compelling governmental reason for testing by one of these novel methods, and that the testing is minimally invasive. If these tests are ever added to the federal program and the legal challenges are resolved, employers will have a firm basis for decision making.

With no federal standard in place for testing these biological specimens, all of these methods are currently being used by employers in the private sector in a trial-and-error approach. With an understanding of the technical limitations and weaknesses of each method and an appreciation of the legal risks in a particular employment environment, employers can make sound decisions and enhance their programs with the proper use of these technologies.

12.5 MANAGING PRESCRIPTION DRUGS: THE USE/ABUSE, TREATMENT/RISK ANALYSIS

The focus of workplace drug testing has been on "illegal" drug use and alcohol misuse and abuse. The primary legal justification for focusing the government's drug testing program on illegal drugs has been in large part the concern for public safety. Private employers share this concern and have additional justifications for their interest and involvement in drug testing. What has become evident is that safety, performance, integrity, and addiction are also being significantly affected by the abuse of prescription drugs.

Workplace drug testing programs do not directly address the safety and performance issues of prescription drug abuse. The essence of the federal model for workplace drug testing programs is to establish a policy that prohibits the use of "illegal" drugs, tests for the presence of these drugs, confirms the presence, and then determines through a confidential dialogue with an MRO whether there is an "alternative medical explanation" for the presence of that drug or metabolite. *Alternative medical explanation* essentially means that the drug was obtained legally by means of direct administration by a physician, such as topical cocaine to a broken nose, or by prescription (e.g., codeine). Once it has been established that the drug was obtained legally, the results are reported as negative.

The existing model does not directly address the issue of whether prescription use is appropriate, such as the prescription for Marinol® for complaints like "nervousness," or the donor's multiple prescriptions for stimulants, or the donor's inappropriate dosing of codeine, or even the use of an old prescription.

The DOT and NRC programs have provisions for the MRO to determine whether a safety issue exists and to report it to the employer or to make an assessment. In the DOT program, this happens when the problem comes to the attention of the MRO, usually inadvertently. Many nonregulated employers are following this approach, and expanding on it by expanding the menu of drugs being tested. Typically the expanded drugs include the synthetic narcotics and benzodiazepines.

The challenge for employers (and society) is how to effectively manage the appropriate use of drugs that have abuse potential and present safety and productivity problems. Although the two largest categories of drugs of concern are historically the synthetic narcotics and the benzodiazepines, there are others.

There are a few cases where employers have taken some independent actions. In an interesting case, a large trucking firm automatically disqualified all applicants who reported that they were taking drugs that were on a list compiled by the firm's safety director. Any drug that was identified in the *Physician's Desk Reference* as requiring a warning about driving was added to the list. The court did not find this to be a violation of the Americans with Disabilities Act.

Testing is, however, only a part of the management of prescription drugs. Employers who are involved in safety-sensitive work have to develop comprehensive policies and procedures that require the disclosure of prescription drugs to a designated medical representative, who then can make an independent judgment in conjunction with the treating physician as to the performance of safety-sensitive functions. The disclosure of prescription drug use can be framed in such a way as not to violate the Americans with Disabilities Act, by avoiding adverse employment action and focusing on the safety issues. These procedures are best justified in clearly safety-sensitive environments, and some degree of employee education is also required. (There need not be any requirement to disclose drug use unless the drug has an effect on performance.)

A related issue has been the growing use of the Internet, not only for filling legitimate prescriptions, but also for obtaining questionable prescriptions for controlled substances. A number of MROs have noted that donors who have tested positive for amphetamines claim to have a prescription and to have obtained the amphetamine from an Internet pharmacy. The issue is not that the prescription was filled by an Internet pharmacy, but rather that the prescription itself was issued by a physician over the Internet, related to or referred from the pharmacy site.

The DEA position is that a prerequisite requirement for the issue of a prescription for a controlled substance is a doctor–patient relationship, and that such a relationship must be based on a face-to-face meeting. Unfortunately, in some states a doctor–patient relationship can be created over the telephone, and telemedicine is only expanding the exception to the face-to-face rule. It is these states that license Internet pharmacies, however, not the DEA.

The problem of Internet prescription writing is not unmanageable for the MRO. Under the DOT regulations the donor has the obligation (burden) to present sufficient evidence to the MRO that his or her drug use that resulted in a confirmed positive laboratory result was obtained by prescription. In the typical Internet case, the donor often has a hard time determining who the prescribing doctor is and establishing that he or she is a licensed physician, or establishing that the drug was dispensed in accordance with federal and state laws. It is highly unlikely that, if there is a physician involved, the physician can establish to whom he or she wrote the prescription. The Internet is anonymous. In these cases the donor cannot meet the minimal requirement of identifying the physician, and the physician cannot establish for whom the prescription was written. In the absence of a face-to-face meeting and any meaningful diagnosis, these cases are typically reported to employers as presenting a safety problem.

12.6 MEDICAL MARIJUANA

Medical marijuana is not a prescription drug, but its use has been decriminalized in 11 states. Most of the state medical marijuana acts list the medical conditions for which the drug can be used and require the "recommendation" of a physician. Many states also have formal patient registries and other conditions of use. In the U.S., the federal government and the states share jurisdiction in respect to the control of legal and illegal drugs. There is usually symmetry in this dual jurisdictional approach, but not always, as seen in the controversy over medical marijuana.

Federal policy has been consistent in prohibiting MROs from accepting medical marijuana as a valid explanation for a positive THC test. Synthetic versions of THC (such as Marinol®, a Schedule III controlled drug) are, however, acceptable. The challenge has been for employers operating in states with medical marijuana acts to develop policies that are legally enforceable. Much of the legal uncertainty has been resolved by the courts.

The most significant court decision in this area is the well-publicized case of *Raich v. Ashcroft*. In *Raich,* the users and growers of marijuana for medical purposes under the California Compassionate Use Act sought a declaration that the Federal Controlled Substances Act (CSA) was unconstitutional as applied to them, in that this federal law is based on the commerce clause, and that growing marijuana legally for one's own use is not "commerce." The plaintiffs prevailed on this argument, which was then appealed to the Supreme Court.

In November 2004, the U.S. Supreme Court heard oral arguments in *Raich v. Ashcroft* (retitled *Raich v. Gonzales*). On June 14, 2005, the Supreme Court released its decision. The Court upheld the federal government's constitutional authority to prosecute the possession, sale, and manufacture of marijuana, regardless of conflicting state law.[22]

The government's brief in the *Raich* case stated:

> The "home-grown" manufacturing, free distribution, and possession of controlled substances, whether for recreational or purported medicinal purposes, also pose an appreciable risk of diversion to others for further drug use or distribution, a result that further swells the illicit market. Local users may ultimately sell or divert the drug to others (for instance, should their production yield exceed their purported needs or should additional funds be required to finance their drug production or other activities).[23]

Justice Souter noted in oral arguments that there could be 100,000 medical marijuana patients just in California, based on the number of cancer/chemo patients. No one disagreed. He noted that carving out an exception for patients in the absence of any other control is *de facto* decontrol. Justice Souter's observation is correct, but the decision in this case, while maintaining the *status quo,* did not end the matter.

The Supreme Court decision did make it easier for employers in California and other states with medical marijuana acts to prohibit the use of native marijuana in any form, particularly for safety-sensitive jobs. But *Raich* did not deal directly with employment issues, and many employers remained worried over possible claims of disability, discrimination, or wrongful termination. Those claims could quite easily be tried in a sympathetic state court with a sympathetic jury and a sympathetic sick employee using medical marijuana.

Employers' concerns over disability claims have been greatly reduced with a significant decision by the Oregon Supreme Court. That watershed case is *Washburn v. Columbia Forest Products,* which was decided in May 2006. The issue that was before the Oregon Supreme Court was whether the Oregon State Disabilities Act (which tracks with the Americans with Disabilities Act) requires an employer to make a disability-related accommodation for an employee who uses marijuana for medical purposes.

The plaintiff, Robert Washburn, was an employee of Columbia Forest Products, Inc. He was also a medical marijuana recipient who regularly used the drug before going to bed to counteract leg spasms that otherwise would keep him awake. After he tested positive for marijuana use, Columbia terminated his employment. Washburn brought action against Columbia, alleging a violation of state prohibitions against disability-related discrimination in the workplace. The trial court granted summary judgment for Columbia, holding, in part, that Washburn was not "disabled" under the pertinent Oregon statutes. The court of appeals disagreed with that conclusion and held that Columbia's summary judgment motion should not have been granted.[24]

Upon review, the Oregon Supreme Court concluded that Washburn was not "disabled." The court followed the U.S. Supreme Court's line of reasoning in previous cases that hold if a drug or corrective action corrects or mitigates the disability, the person is not "disabled" and there is no requirement to make any further accommodation.[25] Setting aside the somewhat circular reasoning of the courts, this decision provides solace to employers in states with medical marijuana laws.[26]

A separate concurring opinion in the case makes the more salient point that, although a state may choose to exempt medical marijuana users from the reach of state criminal law, it is not reasonable to require an employer to accommodate what federal law specifically prohibits.

Although there are employers in jurisdictions where medical marijuana has been excluded from state criminal law sanctions who accept medical marijuana as a legitimate excuse, these employers are making the accommodations more on the basis of employer–employee relations and community standards.

For MROs, the claim of medical marijuana is common in states with medical marijuana laws. MROs recognize that legitimate cases are mixed in with a significant number of bogus claims. Recommendations from physicians to use medical marijuana are not hard to obtain.

Effective management of medical marijuana in non-regulated, non-safety-sensitive employment situations requires some degree of collaboration between MROs and employers, and the establishment of sound employer policies on what is and what is not acceptable.

12.7 SCREEN-ONLY DRUG TESTING

On-site drug testing technology has progressed and diversified to the point where it appears that a broader definition of the term "on-site" is needed. On-site testing initially referred to bench-top automated immunoassay devices used outside of a formal laboratory setting. Over time, as immunoassay methods progressed, instrumental measurement of the immunoassay reaction could be replaced by visual inspection. Immunoassay impregnated on plastic cards became the new "on-site" technology, and the term "non-instrumented on-site testing" was adopted. A variety of different designs exists, from bottles with the immunoassay integrated into the side, to cards that are simply dipped into a urine specimen. Most of the technology is designed to produce a line when the specimen is negative for the specified drug, and the absence of a line indicates a "presumptive" positive test. Most current strip technology performs equally to laboratory screening in identifying negative tests. But like the laboratory-based immunoassay screening, it is not 100% conclusive of the presence of a prohibited drug.

Further, the distinction between instrumented and non-instrumented immunoassay on-site testing is also dissolving. Handheld instrumental devices are in development that will provide an "on-site" analysis of oral fluids and urine. Visual interpretation of a line or no line will be replaced by an automatic report or a simple display. Another innovation is the integration of an on-site instrument that objectively reads the immunoassay results and transmits the results over the Internet to a centralized facility. The computer at the central facility analyzes the results according to a software program and essentially determines whether the specimen should be sent to a laboratory for further testing (either as a non-negative or a random quality control). The final results are then reported to an MRO. This type of approach is a hybrid of centralized laboratory management and local collection and screening. The first such product, called *eScreen*, is already being marketed.

The open question in the utilization of inexpensive immunoassays is to what degree all of the safeguards of training, quality control, and technical management, which are required and used in a certified forensic drug testing laboratory, are applicable to on-site testing, and whether they are necessary.

It should go without saying that existing confirmatory testing requirements will continue to be required in the event that on-site technology becomes incorporated into federally mandated drug testing programs. Confirmatory testing is one of the essential constitutional requirements of federally mandated testing. In workplace drug testing, confirmatory testing has become synonymous with GC/MS technology. But in a broader sense, confirmatory tests can be conducted with LC/MS or MS/MS or comparable instrumentation. The essential element of confirmatory testing is the independent definitive nature of the analysis. It is the true technical foundation upon which sound decisions affecting employment can be made.

All manufacturers and distributors of on-site products are in the chorus recommending GC/MS confirmation and MRO review. However, it is not a foregone conclusion that either of these safeguards will be automatically included in private sector workplace drug testing programs, in the absence of a state law requiring them. Most employers do not have a legal duty or mandate to require confirmation testing, and the economics and apparent simplicity of screen-only results are seductive.

The employment-at-will doctrine is a pillar of labor and employment law in the U.S. The doctrine essentially states that either party to an employment relationship, the employer or the employee, can terminate the relationship for any reason, or for no reason. Thus, it is argued that an employer may terminate an employee based on a screen-only result, but it is an ill-advised practice. First, it is highly unlikely that a screen-only result will be persuasive in any employment proceeding, whether it is an unemployment security hearing, labor arbitration, or a workers' compensation hearing. Further, the employer's publication of a screen-only result to another employer may raise the issue of defamation. This is because for all practical purposes, 99% accuracy is simply not good enough. All screen-positives claim to be that 1%, and the law allows them to make that claim.

A collateral rule to the employment-at-will doctrine is that an employer is free to hire or not to hire applicants based on the employer's assessments, as long as the hiring criteria and decisions are not discriminatory in nature. Thus for many employers the use of a screening device for applicants that produces false positives does not appear to present any significant legal repercussions.

There is essentially no "right" to any job. Applicant screening is the target market for many on-site testing devices. It is a very large market, and many employers do not perceive a need for confirmatory testing or for MROs. In the absence of any state prohibition against screen-only testing of applicants, the argument can be made that the employer who does screen-only testing could run afoul of the Americans with Disabilities Act. Applicants who have been denied a job without confirmation testing may bring an action claiming that they were discriminated against because the employer thought that they were illegal drug users and have a disability, and they do not use illegal drugs and do not have a disability. Although this argument may technically be true, it is not a strong discouragement of screen-only practices.

The broader danger of widespread screen-only practices will be the undermining of the public's acceptance of drug testing. The unwritten social pact is that drug testing is accurate and reliable, and that there will be no adverse impact on an individual who is not currently engaging in illegal drug use.

The open question is whether the incident rate of litigation will continue to remain low, or whether the industry will be faced with increased risk exposure and legal costs. It will not take too many cases for service providers to withdraw from this market.

NOTES

1. The "rules" of workplace drug testing vary considerably and are dependent on such things as who is being tested, why they are being tested, who is doing the testing, and where the test is being collected and performed. An employer in Maine must have a comprehensive program with all the safeguards in place approved by the state before testing, while the same employer in a state with no drug testing law can screen an applicant's urine with an unapproved on-site immunoassay device with impunity. There has been an ongoing process of state legislation that has generated hundreds of statutes that prescribe what is allowed in a particular state and what is not allowed. Today it is nearly impossible for a company to have a uniform drug and alcohol policy and procedure that can be rolled out nationally. This will be a continuing challenge as employers begin to adopt other testing methods.

2. *See,* e.g., *B & F Eng'g, Inc. v. Cotroneo,* 309 Ark. 175, 830 S.W.2d 835 (1992). The court noted that the negligence of impaired transportation workers, such as flight mechanics, bus drivers, or railroad switchmen, can cost an employer millions. *Mulroy v. Olberding,* 29 Kan. App. 2d 757, 30 P.3d 1050 (2001). *But see Beckendorf v. Simmons,* 539 S.W.2d 31 (Tenn. 1976). When the employer had repeatedly forbidden the employee to drive the vehicle, there was no liability.

3. In most collective bargaining agreements, the employer has agreed to provide assistance and a SAP evaluation and return-to-duty requirements for transgressions of the substance abuse policy.

4. The amount of morphine in the specimen was around 600 ng/ml, which is quite consistent with poppy seed ingestion.

5. *SmithKline Beecham Corp. v. Doe.* 903 S.W.2d 347 (Tex. 1995).

6. The following are the cases that reached the appellate level. Drug testing laboratory owes persons tested a duty to perform its services with reasonable care: *Willis v. Roche Biomedical Lab.,* 21 F.3d 1368, 1372-1375 (5th Cir. 1994); *Stinson v. Physicians Immediate Care, Ltd.,* 269 Ill. App.3d 659, 207 Ill. Dec. 96, 646 N.E.2d 930, 932-934 (1995). Laboratory owes prospective employee a duty not to contaminate sample and report a false result: *Nehrenz v. Dunn,* 593 So.2d 915, 917-918 (La.Ct.App. 1992). Laboratory owes employee a duty to perform test in a competent manner: *Elliott v. Laboratory Specialists,* 588 So.2d 175, 176 (La.Ct.App. 1991); writ denied, 592 So.2d 415 (La. 1992). Laboratory owes employee a duty to perform test in a scientifically reasonable manner: *Lewis v. Aluminum Co. of Am.,* 588 So.2d 167, 170 (La.Ct.App. 1991); writ denied, 592 So.2d 411 (La. 1992). Laboratory owes employee a duty to perform tests in a competent, non-negligent manner: See *Herbert v. Placid Ref. Co.,* 564 So.2d 371, 374 (La.Ct.App. 1990).

7. *Harvey J. Duncan, a/k/a Jim Duncan, Appellant (Plaintiff) v. Afton, Inc.,* a Tennessee corporation, d/b/a Healthcomp Evaluation Services Corporation, d/b/a National Ameritest, a/k/a Ameritest; and Leigh Ann Shears, an individual, Appellees (Defendants). No. 99-24. Supreme Court of Wyoming. Nov. 30, 1999. (Cite as: 1999 WL 1073434 (Wyo.).)

8. Claims have been made against MROs, and there have been significant settlements.

9. The general legal duty to remove individuals from safety-sensitive positions is found in 49 C.F.R. 40.23. The prohibition against removing an individual prior to MRO verification is found in 49 C.F.R. 40.21.

10. Didrex is a prescription drug that will cause a positive methamphetamine result. Adipex is a prescription drug that is used as an appetite suppressant; it will not cause a methamphetamine result.

11. It is worth noting that although the actual quantitative value for the methamphetamine was never reported, the initial laboratory result to the MRO reported the presence of methamphetamine only. This means that the amphetamine level was under 500 ng/ml, and such a laboratory result is consistent with inhaler use.

12. Such a practice of universal direct observation of miturition (the process of emptying the bladder), or even the suggestion of doing so, would quickly erode a great deal of public support for drug testing. It may also be viewed as unnecessarily invasive, and trigger various challenges to mandatory drug testing as being an unreasonable invasion of privacy. In fact, many state drug-free workplace statutes require privacy in providing urine specimens. DOT regulations are quite specific on this, and a number of courts have suggested that an observed collection of urine in the absence of individualized suspicion could be viewed as the tort of common law invasion of privacy (e.g., *Borse v. Piece Goods,* 963 F.2d 611 (3d Cir. 1992), applying Pennsylvania law). In *Borse,* the plaintiff was discharged for refusing to submit to urinalysis screening and to personal property searches at the workplace. The federal appellate court predicted that, if faced with the issue, the Pennsylvania Supreme Court would find a public policy exception to the employment-at-will doctrine based on the common law tort of invasion of privacy if the employer was going to require Mrs. Borse to provide urine under observed collection (in the absence of individualized suspicion).

13. Department of Health and Human Services Substance Abuse and Mental Health Services Administration, *Mandatory Guidelines for Federal Workplace Drug Testing Programs,* Tuesday, April 13, 2004 (69 FR 19644-01).

14. Edgell, K., Laboratory survey of adulterant testing, *MROAlert,* Volume XVI, No. 10 (December 2005).

15. U.S. Department of Transportation. Procedures for Transportation Workplace Drug and Alcohol Testing Programs, Notice of Proposed Rule Making, October 31, 2005 (70 FR 62276).

16. Federal Aviation Administration, Office of Aviation Medicine, *Workplace Urine Specimen Validity Testing Colloquium,* Tampa, Florida (February 4–6, 2003).

17. 1998 DOT regulations, 49 C.F.R. Subtitle A, §40.25(f)(22)(ii).

18. See also 2001 DOT §40.355: "What limitations apply to the activities of service agents? As a service agent, you are subject to the following limitations concerning your activities in the DOT drug and alcohol testing program. (a) You must not require an employee to sign a consent, release, waiver of liability, or indemnification agreement with respect to any part of the drug or alcohol testing process covered by this part (including, but not limited to, collections, laboratory testing, MRO, and SAP services). No one may do so on behalf of a service agent."

19. GAO-05-653T: "Drug Tests: Products to Defraud Drug Use Screening Tests Are Widely Available."

20. Department of Health and Human Services, Substance Abuse and Mental Health Services Administration. Proposed Revisions to Mandatory Guidelines for Federal Workplace Drug Testing Programs, April 13, 2004 (69 FR 19673-01).

21. On-site oral fluid testing devices are available but are essentially ineffective at identifying the parent drug THC in specimens. THC is the target for oral fluid testing since the amount of metabolite (THC-COOH) found in urine is found in very small quantities in the mouth.

22. *Alberto R. Gonzales, Attorney General, et al., Petitioners, v. Angel McClary Raich et al.* No. 03-1454. Supreme Court of the United States. Decided June 6, 2005. 125 S.Ct. 2195.

23. In the Supreme Court of the United States, *John D. Ashcroft, Attorney General, et al., Petitioners v. Angel McClary Raich, et al.* On Writ of Certiorari to the United States Court of Appeals for the Ninth Circuit, No. 03-1454. Brief for the Petitioners, Paul D. Clement, Acting Solicitor General.

24. *Washburn v. Columbia Forest Products, Inc.,* 197 Or App 104, 104 P3d 609 (2005).

25. The U.S. Supreme Court had held that a person is not disabled under federal disability law if a mitigating measure will alleviate an otherwise substantial limitation to a major life activity. *See, e.g., Sutton v. United Airlines, Inc.,* 527 US 471, 119 S Ct 2139, 144 L Ed 2d 450 (1999).

26. What appears circular in applying the mitigation rule to the use of medical marijuana is that the decision essentially prohibits the employee from taking advantage of the very medication that alleviates his purported disability.

Index

A

Absorption, 99
Abuscreen ONLINE, 59
Accessioning, 16, 28
Accrediting organizations, 161–163
Accuracy, confirmatory testing, 66–67
Accu-sorb, 88, 89, 91
Administrative cutoffs, 159
Administrative functions, MRO, 150
Adulteration
    laboratory testing, 28
    legal issues, 175, 177
    specimen validity testing, 136–140
Afton, Inc., 172
Alcohol
    fitness-for-duty proposed amendments, 39–40
    testing, 30–32, *31*
    violation, 26
Alternative matrices, analytical approaches
    fundamentals, 82, 95
    hair, 82–87
    legal issues, 175
    oral fluid, 88–91
    sweat, 92–94
Alternative matrices, test result interpretation
    absorption, 99
    analytical considerations, 106, 108
    blood, 100–102, *101*
    chemical considerations, 106, *107,* 108
    clearance, 99–100
    detection time course, 108, *111*
    disposition of drugs, 98–100
    distribution, 99
    drug disposition, 98–100
    elimination, 99–100
    exposure circumstances, 98, *99*
    fundamentals, 97–98
    guidance in interpretation, 111–112, *113*
    hair, 104–106, *106*
    interpretation, 108–112, *113*
    multiple specimen testing, 111
    oral fluid, 103, *103*
    pharmacokinetic considerations, 106, *108, 109–110*
    physiologic considerations, 100–106
    sweat, 104, *104*
    time course of detection, 108, *111*
    urine, 102, *102*
Alternative medical explanation, 179
Amphetamines
    derivative selection, 72–73
    hair, 86
    immunoassay testing, 60–62
    ion selection, 75
    oral fluid, 90
Analytical considerations and approaches
    accuracy, 66–67
    alternative matrix test results interpretation, 106, 108
    amphetamines, 60–62, 72–73, 75
    analytical issues, 72–76
    assay calibration, 69
    barbiturates, 64, 75–76
    benzodiazepines, 64, 73–74, 76
    calibrators, 69–71
    cannabinoids, 62
    carryover, 68
    chromatographic performance, 69
    cocaine, 62–63
    confirmatory testing, *65,* 65–76
    controls, 69–71
    criteria, positive test results, 71–72
    data review, 71–72
    derivative selection, 72–74, *73*
    enzyme immunoassay, 58–59
    evaluation, 70–71
    fluorescence polarization immunoassay, 59
    fundamentals, 56, *57*
    HHS-regulated drugs, 60–64
    immunoassay testing, 57–65, *58*
    interference studies, 67–68
    internal standard selection, 74–75
    ion selection, 75–76
    linearity, 67
    methadone, 64, 76
    methaqualone, 65
    method validation, 66–68
    on-site drug testing, 60
    opiates, 63–64, 74
    particle immunoassay, 59–60
    phencyclidine, 64, 76
    precision, 66–67
    propoxyphene, 65, 76
    qualitative criteria, 71